Basic Superfluids

The Taylor & Francis *Masters Series in Physics and Astronomy*
Series Editor: David S. Betts
Department of Physics and Astronomy, University of Sussex, Brighton, UK

Core Electrodynamics
Sandra C. Chapman
0–7484–0622–0 (PB)
0–7484–0623–9 (HB)

Atomic and Molecular Clusters
Roy L. Johnston
0–7484–0931–9 (PB)
0–7484–0930–0 (HB)

Quantum Theory of Solids
Eoin O'Reilly
0–7484–0627–1 (PB)
0–7484–0628–X (HB)

Basic Superfluids
Tony Guénault
0–7484–0892–4 (PB)
0–7484–0891–6 (HB)

Basic Superfluids

Tony Guénault

Taylor & Francis
Taylor & Francis Group

LONDON AND NEW YORK

First published 2003
by Taylor & Francis
11 New Fetter Lane, London EC4P 4EE

Simultaneously published in the USA and Canada
by Taylor & Francis Inc,
29 West 35th Street, New York, NY 10001

Taylor & Francis is an imprint of the Taylor & Francis Group

Typeset in Palatino by
Integra Software Services Pvt. Ltd, Pondicherry, India
Printed and bound in Great Britain by
MPG Books Ltd, Bodmin

British Library Cataloguing in Publication Data
A catalogue record for this book is available from the British Library

Library of Congress Cataloging in Publication Data
A catalog record for this book has been requested

ISBN 0–7484–08916 (hbk)
     0–7484–08924 (pbk)

Contents

Figures

Preface

This book has come out of a desire to provide something readable which covers the whole topic of superfluids, one of great fascination to me. In the book, I assume a knowledge base corresponding to most core (second or third year) physics degree courses, but especially those in thermal and statistical physics and in quantum mechanics. I hope that this tour through the world of superfluids will inform and up-date many physics graduates, including those involved in teaching physics. Specifically, the book is aimed to fill the needs of final-year students who are doing courses involving low temperature physics. It should also be of particular interest to research students starting their work in allied fields of study.

I have not attempted to be comprehensive, or indeed to be particularly scholarly in every last detail, although I hope that I have not been misleading in any respect. The book concentrates on the behaviour of condensed matter in equilibrium, that is on topics traditional to low temperature physics, so that I have not included extended treatments of Bose–Einstein condensation of cold alkali metal atoms or of possible superfluidity in neutron stars. The longest chapter is the final one on superfluid helium-3, for two reasons. One is a perceived need for an elementary introduction to this fascinating, rather new and somewhat complicated superfluid. The other is that the topic has occupied much of my own working life for the past 20 years, following an earlier career in metal physics including superconductivity.

Writing a book on superfluids has been greatly helped by many inter-actions with my colleagues at Lancaster, George Pickett, Shaun Fisher and Ian Bradley. Their insight, friendship and generous practical assist-ance have been, and still are, of much value to me. Having said that, any inaccuracies in the text are of course entirely my own work. But above all, I have received constant encouragement and support from my wife, Joan, and this book is written for her.

Tony Guénault
Lancaster, February 2002

Chapter 1

What happens at low temperatures?

Low temperature physicists have goals which are common to most human beings. They are looking for a quiet life! Perhaps regrettably, the state of perfect order and peace which they seek is not so much in their heads as in the physical world outside them. Fortunately, in this physical world, it turns out that the calming down of thermal vibrations at low temperatures does not merely bring a grey dullness and sameness to the properties of every form of matter. Rather, it allows the emergence of a whole range of new states of matter which unfold a surprising richness and excitement. Among these new states are the "superfluids".

Superfluidity has been described as the jewel in the crown of low temperature physics. Superfluid states are characterised by the dominant influence of quantum mechanics on the large-scale thermal properties of the substance in question. As we shall see, superfluid states are observed in many systems. Examples are found in the superconductivity of many metals and some near-metals, in the extraordinary properties of the helium liquids, in neutron stars and in the behaviour of gaseous assemblies of cold alkali metal atoms.

1.1 ENTROPY, ORDERING AND THE THIRD LAW

What happens at a low enough temperature is that anything which is in thermal equilibrium must become ordered, ordered in the sense of organised. In other words it has zero entropy. That in essence is the content of the Third Law of Thermodynamics.

1.1.1 Ordering in a simple substance

Let us take nitrogen as a typical substance, one with few complications and in which no superfluid state is known. (We conveniently avoid remembering that N_2 molecules have internal structure, since that does not affect the argument here!) How does ordering take place as our nitrogen is cooled towards the absolute zero of temperature?

At room temperature nitrogen is of course a gas, which is a highly disordered state of matter. The gas has a high entropy. Furthermore when the temperature is increased, the molecules move even faster, the gas expands at constant pressure and it becomes more dilute and more disorganised. The entropy thus increases further. When the gas is cooled below room temperature the reverse happens. The molecules slow down and the volume occupied by the gas at constant pressure decreases, roughly according to the ideal gas law,

$$PV = RT. \tag{1.1}$$

Correspondingly the entropy of the gas decreases.

However this ideal gas behaviour does not continue indefinitely as the temperature T is lowered. Instead, the interactions between molecules take over as their kinetic energy (of order $k_B T$ where k_B is Boltzmann's constant) becomes smaller and their spacing is reduced. At the boiling point (around 77 K for nitrogen) a sudden phase change occurs and the substance condenses into a liquid. The liquid is a phase of higher orderliness (lower entropy) because the density is so much higher than in the gas phase. The latent heat L of the phase transition speaks directly of this entropy reduction ΔS since $L = T \Delta S$.

In the liquid phase, the nitrogen molecules remain highly mobile with kinetic energy still of order $k_B T$. As T is reduced further, there comes a point where the liquid freezes (about 63 K for nitrogen) to become a solid. Another, latent heat demonstrates the advent of this even more ordered phase, in which molecules are effectively localised on to well-defined lattice sites.

Further reduction in temperature produces no further changes of phase. The quantum ground state of the solid corresponds to all molecules being located at their lattice sites as securely as possible and having the minimum possible vibrational energy. It is as well to recall that this minimum energy is not zero, however, because of Heisenberg's uncertainty principle. For example, a simple (one-dimensional) harmonic oscillator has a minimum ("zero-point") energy of $1/2\, h\nu$ where h is Planck's constant and ν the classical oscillator frequency, since the oscillating particle cannot simultaneously have its position certain and its momentum definitely zero. This type of consideration affects each degree of freedom in the solid.

Above absolute zero, the solid has vibrational energy usually described as a gas of phonons. As the absolute zero is approached, these thermal phonons are simply frozen out, a situation analogous to the freezing out of thermal photons, i.e. black-body radiation, in an "empty" box containing only electromagnetic radiation. This is possible because phonons (and photons) are particles whose number is not conserved, so that they can

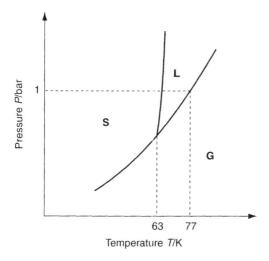

Figure 1.1 The phase diagram of nitrogen (schematic). Note the triple point at 63 K and the normal boiling point at 77 K.

simply disappear as T is reduced without any problem. In this respect, the statistical physics of such a "phony" gas (pun intended!) differs dramatically from that of a real gas, as we shall see below.

The phase diagram of nitrogen in Figure 1.1 illustrates its ordering behaviour and the existence of the three common phases: gas, liquid and solid. Note that the existence of a triple point ensures that at absolute zero the only possible phase is the solid. One can note that it is only because the triple point pressure is below one atmosphere (1 bar) that we experience the liquid phase at all, under common conditions. Contrast carbon dioxide, triple point pressure 5 bar, which we meet as a gas or as a cold solid, "dry ice", or recall that nitrogen forms the dry ice of Triton (Neptune's largest moon) which has an atmospheric pressure of a minute 1.5×10^{-5} bar and a temperature of 38 K.

1.1.2 Ordering in not so simple substances

In practice, ordering at low temperatures can of course be more complex than that of the above. There are two major types of complications which have been brushed under the carpet in the discussion so far, even before we get on to the topic of superfluids.

The first complication to be recognised is that there are several import-ant contributions to the thermal properties of many substances, other than the simple molecular motion discussed above. Hence there are other entropy contributions to be considered in practice. For example, in many materials we do not have essentially isolated molecules, but in the solid and/or liquid states the proximity of atoms leads to electronic overlap and hence to non-localised electronic motion, often giving rise to metallic or semiconducting properties. Indeed it is the existence of such an electron gas at the lowest temperature which leaves the way open for superconductivity, as we shall see. Then, secondly, there is the existence of degrees of freedom associated with electronic and nuclear spin and with orbital motion of electrons around nuclei, all of which lead to magnetic properties. The consequent ordering of electronic or nuclear spins at low temperatures gives a variety of important phenomena. Thirdly, there are structural phase transitions in many solids, arising from the subtle way in which interatomic interactions vary with separation and direction. Fourthly, we can have contributions from defects and vacancies.

As if that is not enough to consider in real life, there is another type of complication entirely. This relates to whether the substance under discussion is in fact in thermal equilibrium. This is a matter of time scales. At low temperatures, the time scale for thermal equilibrium to be achieved can become effectively infinite for some types of reordering, particularly those involving the movement of molecules. Hence disorder is frozen in as the temperature is lowered. A classic example is that of a glass, where even at room temperature the molecules form a disordered and unpredictable array rather than a regular crystalline structure. And the glassy structure once formed remains in the same state for a very long time. Thus the true quantum mechanical ground state of the system is never accessible.

In contrast, in the situations where superfluids are found, the ground state is certainly accessible. Indeed it is the coherent quantum nature of this state which plays a central role in the observable properties of the substance. But before discussing this further, we take a closer look at the Third Law and its implications.

1.2   THE THIRD LAW OF THERMODYNAMICS AND ORDERING

The Third Law of Thermodynamics, originally called the Nernst Heat Theorem after it was formulated by Nernst in 1906, was a matter of considerable controversy until it was clarified by Sir Francis Simon in the 1930s.[1] Stated in its strongest form the law simply says that "the entropy of any substance must go to zero as the absolute zero of temperature is

---

1 For a fuller discussion see the book "Entropy and its Physical Meaning" (Taylor and Francis 1996) by J. S. Dugdale, once a research student in Oxford with Simon.

reached". The difficulty of this formulation is obvious from the discussion of the previous section. It is not going to be either true or very useful if the substance never reaches its true thermodynamic equilibrium state.

The entropy S of a substance can be expressed in terms of its possible quantum states (so-called microstates) as

$$S = k_B \ln \Omega, \tag{1.2}$$

where Ω is the number of possible quantum states which the whole system can achieve. This expression, first formulated by Boltzmann and extended by Planck, gives a valuable statistical interpretation of the meaning of entropy. Zero entropy thus implies that the system must have $\Omega = 1$, i.e. that it must certainly have entered its ground state. Hence another way of looking at the Third Law is that it states that "any substance must reach its ground state as it approaches $T = 0$".

1.2.1 Is equilibrium reached?

A modern view of the Third Law includes a consideration of equilibrium times. The key, as recognised by Simon, is to recognise explicitly that there are many independent contributions to the entropy of a typical substance. In other words, there are many quantum numbers which must be specified to describe the quantum state of the system. Take, for example, a block of copper metal. This has (at least) the following possible contributions to its entropy:

1   Lattice vibrations, phonons.
2   Conduction electrons.
3   Nuclear spins, since both isotopic constituents ^{63}Cu and ^{65}Cu have spin 3/2.
4   Isotopic disorder from arrangements in the crystal lattice of ^{63}Cu and ^{65}Cu.
5   Crystalline imperfections (grain boundaries, dislocations, vacancies).

Now the point is that, as a broad approximation, one can divide such contributions to the disorder of a substance into two classes. First, there are those which come to equilibrium under the relevant experimental conditions of temperature and time scale. These contributions obey the Third Law in a limited but useful form: "As absolute zero is approached, any contribution to entropy which comes to equilibrium must become zero." Items 1, 2 and 3 in the list above are all firmly in this category. Aside, we may note that often one should also think about thermal equilibrium between contributions as well as within a contribution. In the case of copper at microkelvin temperatures and below, the nuclear spins (item 3) are only weakly coupled to the electrons and lattice (items 1 and 2), even though they achieve rapid thermalisation with themselves. Therefore, on the appropriate time scale, it can be thermodynamically realistic to talk about the spins having one equilibrium temperature, whereas the lattice and electrons have another. This is an important idea in the understanding of nuclear cooling to be discussed later.

Secondly, there are those contributions to disorder in which no changes are made under the experimental conditions. One may describe such entropy contributions as "frozen in". Items 4 and 5 for copper are examples of this, item 4 at all temperatures, item 5 at room temperature and below. Isotopic disorder (item 4) is an interesting and clear-cut example. If we are doing our experiments in a gravitational field (as one usually does without the support of the International Space Agency!), then the ground state should have all the ^{65}Cu atoms at the bottom of the sample and the lighter ^{63}Cu at the top. This would be the state of lowest potential energy. But such isotopic separation does not take place simply because the gravitational potential energy differences are so small, and the energy barrier to interchange atoms is very high. Hence the isotopic disorder which existed when the metal was formed persists for all time. Similarly, many structural defects (item 5) are frozen in at a low enough temperature. Again this arises since a significant energy barrier (ΔE, say) often needs to be overcome for a structural rearrangement to be made. As $k_B T$ becomes lower than ΔE the probability of any change is frozen out by the usual Boltzmann factor $\exp(-\Delta E/k_B T)$.

This last simple expression gives the key to our use of the Third Law. When low temperatures are being considered, it is clear that any form of reordering which does not have an energy barrier will follow the Third Law – the single thermodynamic ground state will be approached as the temperature is lowered. On the other hand, if there is an energy barrier, then as soon as $k_B T$ becomes significantly less than ΔE, then an overwhelming freeze-out takes place. The probability for any change, linked to the exponential Boltzmann factor, becomes negligibly small. Hence the system remains locked in to a single metastable state, the properties of which depend on the history of the sample. But what is important to appreciate is that our system really does stay frozen into this one metastable state, so that this form of entropy contribution plays no part in the observed thermodynamic properties of the substance. Hence even these contributions obey a weaker (but useful) form of the Third Law, namely "All entropy changes must tend to zero as absolute zero is approached."

1.2.2   Consequences of the Third Law

So just what are these useful consequences of the Third Law? Suppose that for the most part we are talking about degrees of freedom of a system which remain in thermal equilibrium to the lowest temperatures. A few such consequences follow:

1   The ground state of the system can not be degenerate. Remembering that $S = k_B \ln \Omega$, if it were degenerate, then the system would have a choice of states and hence the entropy would be greater than zero. As noted earlier, $S = 0$ for that entropy contribution implies that $\Omega = 1$.
2   A special case of this idea indicates that there is no such thing as an ideal paramagnet!! When the external magnetic field applied to an assembly of localised spins is turned off, the Zeeman splitting just cannot go to zero. Physically, this means that interactions in an assembly of spins must always take over below some temperature. We shall return to this point later in considering nuclear cooling (see Section 3.1.2).
3   Related is a common, if bombastic, statement of the Third Law, which can be derived from the above. This is that "It is impossible to reach the absolute zero." As we shall see in Chapter 3, any cooling method relies on a working substance whose entropy can be changed at the temperature of interest by changing a control parameter (for instance pressure or magnetic field). At $T = 0$ our third law tells us that this is impossible. In fact, this is impossible in the case of frozen-in disorder, as well as in the true equilibrium situation, since that is the direct implication of "freezing in".
4   If you love phase diagrams (see Figure 1.1 again), then a further interesting consequence follows from the Clausius–Clapeyron equation, namely that the slope dP/dT of any line in the diagram must be zero if it extends to $T = 0$. The Clausius–Clapeyron equation tells us that for the two phases to be in equilibrium

$$\frac{dP}{dT} = \frac{\Delta S}{\Delta V} \tag{1.3}$$

where ΔS is the entropy difference between the two phases and ΔV the volume difference. Since ΔV is non-zero, the Third Law demands that ΔS must be zero at $T = 0$, and hence the result.
5   Finally, since dS/dT is not allowed to do anything exotic, it follows that the heat capacity $(C = TdS/dT)$ must vanish at the absolute zero. Again one may note that this requirement also applies to all separate entropy contributions in the substance, since again any frozen-in metastable disorder has zero contribution.

1.3 ORDERING IN HELIUM

Helium is unique among chemical substances in that the fluid states exist right down to the absolute zero. At room temperature helium is

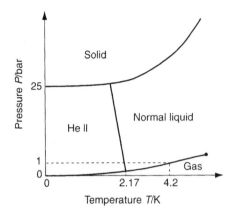

Figure 1.2 The phase diagram of ⁴He. Note the absence of a triple point. The λ-point is at 2.17 K and the boiling point at 4.2 K. Solid only exists at pressures above 25 bar. The gas–liquid line terminates at the critical point.

the most ideal gas, and in its gaseous state, the common isotope ⁴He survives down to around 4.2 K before it liquefies. And below that temperature it never solidifies unless subjected to at least 25 atmospheres of pressure. The phase diagram for ⁴He is shown in Figure 1.2, and should be contrasted with that of nitrogen (Figure 1.1). The connectivity of the diagram is quite different in that there is no triple point in helium.

Fortunately for low temperature physics, ⁴He is not the only available isotope of helium, but ³He is also stable. Helium gas is obtainable as a minority constituent (up to 7% in some cases) from some natural gas sources, the amount in the atmosphere being minute because of the small mass. However this natural helium contains typically only about 1.3 parts per million of ³He. Hence the use of ³He in cryogenics is comparatively recent (around 1960), made possible by the commercialisation of nuclear reactions, since it is a product of tritium decay.

The light isotope ³He has the phase diagram shown in Figure 1.3. The boiling point is now 3.2 K and a pressure of 34 bar is required to produce solid at low temperatures. Again, there is no triple point, so the connectivity of the phase diagram, but not its shape, is similar to that of ⁴He.

All this raises (at least) three questions, each of considerable interest.

1   Why is there no solid, except at high pressure?
2   Why is the detail for ³He different from that for ⁴He?
3   What is the nature of the liquid state as $T=0$ is approached?

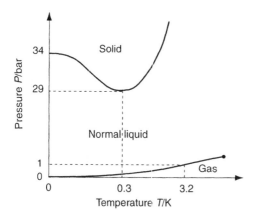

Figure 1.3 The phase diagram of ^{3}He. Again note the absence of a triple point. The solid liquid line has a minimum around 0.3 K. Below this temperature the liquid is more ordered (by Fermi–Dirac statistics) than the solid. Superfluidity only occurs below about 0.002 K, effectively hidden by the y-axis on this scale.

1.3.1 Why is there no solid?

The reason, in outline, can be stated briefly: "Quantum mechanics and the Uncertainty Principle."

As mentioned above in Section 1.1.1, our picture of a solid at absolute zero is of the array of molecules in their lowest vibrational state. But as we pointed out this lowest energy state is not zero. A harmonic oscillator mode of frequency ν has a minimum energy of $1/2\, h\nu$. This is related to Heisenberg's Uncertainty Principle which asserts that a particle cannot have position and momentum both definitely fixed, so that the absolute zero state of a solid cannot be that of each molecule neatly screwed to its lattice site; rather a certain "zero point motion" must exist. In all substances except helium this motion has negligible consequences. However, in helium the atoms are sufficiently loosely bound by the weak van der Waals forces that even this zero point motion is enough to make the solid unstable, i.e. it melts under its own zero point energy. It is only when the solid is stiffened by pressure that the solid can become the stable phase. This is a feature in common between ^{3}He and ^{4}He. However that brings us to the next question.

1.3.2 Why are the solid–liquid lines for ^{3}He and ^{4}He so different?

Another brief answer, again based on quantum mechanics, can be given: "3 is an odd number whilst 4 is even."

This staggering mathematical statement turns out to be of critical importance to all of the properties of ^{3}He and ^{4}He at low temperatures. The liquids are often referred to as examples of "quantum fluids", because of the dominant importance of the wave function symmetry demanded by quantum mechanics. Hence this simple odd/even idea will be the subtext of the next section and indeed much of the rest of the book. At this stage, the only relevant point to note is that ^{4}He has 2 protons and 2 neutrons in the nucleus with 2 electrons around. Each of these spin 1/2 fundamental particles has its pair, so that the total spin of the ^{4}He atom is zero. Hence ^{4}He has no significant magnetic properties. The entropy of the solid derives only from its atomic motion, so that it is extremely small at all relevant temperatures. It is also true that the entropy of liquid ^{4}He below about 2 K is very small. Consequently ΔS for the liquid–solid transition is very small, leading inevitably from the Clausius–Clapeyron equation (equation 1.3) to a very small dP/dT, i.e. to an aggressively horizontal line on the phase diagram (Figure 1.2).

On the other hand ^{3}He (2 protons but only 1 neutron) has a net nuclear spin of 1/2 because of the unpaired neutron. Hence solid ^{3}He is an almost ideal "spin 1/2 solid", with two possible nuclear spin states per atom. This leads to a lot of disorder in the solid ($k_B \ln 2$ per atom) which persists to really quite low temperatures. In fact, it turns out that the spins are sufficiently weakly interacting that the spin disorder is not frozen out until around 1 mK. In contrast, the spin 1/2 property of the liquid ^{3}He atoms means that (to a surprisingly good approximation) it can be described as an ideal Fermi–Dirac gas. At temperatures below its Fermi temperature (about 0.15 K), the exclusion principle is important. The absolute zero ground state of the gas has all one-particle states definitely occupied up to the Fermi energy E_F and definitely unoccupied above it. This definiteness ensures that at $T=0$ we have $\Omega=1$ and hence $S=0$. At modest temperatures, the thermal excitations above this state are restricted by the available thermal energy ($k_B T$), so that only a fraction $k_B T/E_F$ of the fermion gas particles are effective in contributing to the thermal properties. In particular, the entropy and heat capacity C_v of the "gas" are of order $Nk_B (k_B T/E_F)$, linear in temperature.

The entropy curves for liquid and solid ^{3}He are illustrated in Figure 1.4. The astonishing (and unique) result is that the solid below 0.3 K is more disordered than the liquid. Disorder in the liquid is restricted by the Fermi–Dirac statistics. Hence the entropy difference $\Delta S = S_{liquid} - S_{solid}$ between the phases is negative, and the famous Clausius–Clapeyron equation tells us that (since the volumes of the two phases behave in the usual manner) we have dP/dT negative also in this temperature range. Hence the relation between the entropy diagram and the phase diagram (Figure 1.3). It is a remarkable example of the consistency of a thermodynamic argument, almost enough to make you believe the theory!

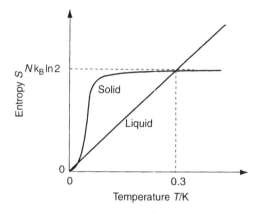

Figure 1.4 Entropy–temperature graph for ^{3}He liquid and solid. The liquid state has the lower entropy below 0.3 K, relating to the phase diagram (Figure 1.3).

Before leaving this topic, we may note that this negative ΔS also implies a negative latent heat, generating cooling when liquid is converted into solid. Hence applying increasing pressure to a ^{3}He liquid–solid mixture, which converts some liquid into solid, provides a practical mechanism for cooling to around 1 mK, below which temperature the spins in the solid order and the solid entropy falls (Figure 1.4). This technique is called "Pomeranchuk cooling" (see Section 3.1.2), and was the method used in 1971 by Lee, Osheroff and Richardson in their Nobel Prize winning discovery of superfluidity in liquid ^{3}He.

1.3.3 What is the nature of the liquid state as $T = 0$ is approached?

This question forms the main topic of this book, so cannot be answered too succinctly. However, the obvious comment is: "Well, the liquid must be in an ordered state, whatever that means!"

Again we must note that the liquid part of the phase diagram for ^{4}He is quite different from that for ^{3}He. The point about 3 being an odd number and 4 an even one turns out to be of major importance. The different symmetry properties for fermions (odd numbers) as opposed to bosons (even numbers) is the root cause of the distinction.

Since a liquid obviously has mobile atoms, it makes sense to start by understanding the nature of an ideal gas of weakly interacting identical particles at low temperatures.

We have discussed the Fermi–Dirac case earlier. The picture of the absolute zero ground state is of one-particle states definitely filled up to the Fermi energy E_F and definitely empty above it. This is a state of zero entropy, but also one of very high energy. This is described by the Pauli exclusion principle. The statistics demand a total wavefunction which is antisymmetric, which means that the wavefunction must change sign for the interchange of coordinates of two identical Fermi particles. Hence no two particles can be in the same state, since this situation would have to be described by a wave function which was minus itself, i.e. zero. In other words, the exclusion principle operates. Hence, to accommodate all the identical particles which make up the gas, states up to a high energy must be occupied. This is "tower block ordering".

The Bose–Einstein gas becomes ordered in an entirely different manner. The wavefunction is now symmetric for interchange and so does not vanish when two particles occupy the same state. There is no exclusion principle. Hence it is unsurprising that the overall ground state of a Bose gas finds all of the particles occupying the same lowest energy state, the one-particle ground state. This is "basement ordering".

But there is more to be said about the ideal Bose gas. A fascinating feature is that as the temperature is lowered, this ordering suddenly starts to come in at a characteristic temperature T_B. Above this "Bose–Einstein condensation" temperature, the gas obeys statistics which is simply a slight modification of the usual Maxwell–Boltzmann statistics of a classical gas. Below T_B, a macroscopic fraction of the particles "condense" into the ground state. This fraction increases from zero at T_B as the temperature is lowered until it equals unity at $T=0$. The important term "condensate" is used to describe the fraction of particles in the ground state, all described by the same (coherent) wavefunction.

These simple ideas, based as they are on an ideal gas at $T=0$, contain only part of the answer to the question about the helium liquids, since the liquids by definition consist not only of fluid, but also of strongly interacting, atoms. We consider the influence of such interactions further in the next section. However, with reference to the phase diagrams of Figures 1.2 and 1.3, we may note that ^{4}He, the boson system, does have a phase transition in the liquid state at a high temperature (about 2.17 K at vapour pressure). Such a transition is believed to be directly related to the Bose–Einstein condensation. On the other hand, there is no such phase transition at similar temperatures in the Fermi system ^{3}He. Instead, it shows the characteristic $(k_B T/E_F)$ properties of an ideal Fermi gas, as noted in the previous section, at temperatures right down to the millikelvin regime, before it finally enters a superfluid state.

1.4 WHAT MAKES A SUPERFLUID?

A necessary condition for the existence of a superfluid state is that the substance forms a condensate. In other words, we require there to be a single coherent quantum state of the whole substance which contains a macroscopic fraction of the particles making up the substance. As we shall discuss in the following section, this actually is not the only requirement for superfluidity; it is not in itself a sufficient condition.

For Bose systems, the requirement of a condensate is met even in an ideal gas. For Fermi systems, rather astonishingly, it is only when interactions are significant that a superfluid can emerge. If there is an attractive interaction between fermions, then it turns out that at a low enough temperature the single particles become paired; and the pairs ape boson behaviour, since two odds make an even.

The interest and excitement of superfluidity will be affirmed in this section with a trip through the history of the Nobel Prize in Physics. The study of matter at low temperatures was initiated by Heike Kamerlingh–Onnes (Nobel Prize 1913) who both liquefied helium and discovered superconductivity. The pioneering work of Pyotr Leonidovich Kapitza was recognised late in life (Nobel Prize 1978 when he was already aged 84). Meanwhile the 1962 Nobel Prize went to Lev Davidovich Landau, a great Russian physicist who made seminal contributions to our understanding of the nature of superfluidity (as outlined later in this chapter) and also to the theory of Fermi liquids.

1.4.1 Examples of Bose systems

The two clear examples of condensed Bose systems are

1   Liquid ^{4}He.
2   Gaseous assemblies of cold atoms.

Liquid ^{4}He, as explained above, is in no way an ideal gas of ^{4}He atoms. The atoms interact strongly, which is of course why the liquid state is formed. Nevertheless, the liquid shows a phase transition, which in fact is more violent in nature than that of the Bose gas. As an indication of the transition there is the so-called λ-transition in the graph of heat capacity of ^{4}He against temperature as shown in Figure 1.5. Also shown for comparison is the calculated heat capacity of an ideal Bose gas of the same density. One can remark that the transition temperature occurs at roughly the "right" temperature, although the detail is quite different. We consider the properties of liquid ^{4}He further in the next section and in Chapter 2.

In recent years, several research groups have found exciting new physics in the properties of cold atoms. The work was recognised by the award of the 1997 Nobel Prize in Physics to Steven Chu, Claude

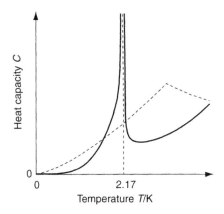

Figure 1.5 Heat capacity of liquid ^{4}He (full line) showing the λ-point anomaly. For comparison, the heat capacity of an ideal Bose gas of similar density is shown (dashed line); this shows a much gentler anomaly at the Bose–Einstein condensation temperature.

Cohen-Tannoudji and William D. Phillips. In a typical experiment, they confine a small amount of an atomic vapour consisting of identical bosons, such as that of the alkali metal rubidium, in a magnetic trap. The atoms are slowed down by suitable laser interactions, to give very low kinetic temperatures; and they may be cooled further by expansion of the trap. It has been demonstrated that the collection of such atoms do indeed undergo Bose–Einstein condensation, and the study of its properties is a fascinating topic of current atomic physics research. As I write this book, the 2001 Nobel Prize in Physics has been awarded to Eric Cornell, Wolfgang Ketterle and Carl Wieman for "the achievement of Bose–Einstein condensation in dilute gases of alkali atoms, and for early fundamental studies of the properties of the condensates". This is a fascinating cross-over area between atomic physics and equilibrium low temperature physics, the topic of this book.

1.4.2   Examples of Fermi systems

Fermi systems which show superfluid states include the following:

1   Electrons in metals.
2   Liquid ^{3}He.
3   Electrons in strange compounds.
4   Highly dense matter in stars.

As stated above, a small attractive interaction between identical Fermi particles can have a dramatic effect on the nature of the ground state. Such an interaction can have quite a subtle cause.

Consider the case of conduction electrons in metals. Clearly one must expect these identical charged particles to repel each other by simple electrostatics. And they do so at short range, although one must recognise that the metal also contains positive ions and is overall neutral. However in some metals, but it seems not in all, another effect holds sway at low enough temperatures. This is the formation of so-called Cooper pairs of electrons, a vital ingredient of the BCS theory of superconductivity for which John Bardeen, Leon N. Cooper and J. Robert Schrieffer were awarded the 1972 Nobel Prize for Physics. A handwaving description of the pairing is as follows. It is based on the simple Coulomb attraction between an electron and a positive ion. One electron passing rapidly through the crystal lattice of the metal causes a short-lived disturbance (a "virtual phonon" in technical language) in the lattice ions. Another electron, in fact one travelling in the opposite direction, can then sense the disturbance left by the first one even though the two electrons are mainly far apart. In the superconducting ground state, states are occupied in these correlated pairs; the pairs may be thought of essentially as bosons, and the ground state as a condensate. It is relevant to note that this Cooper pairing is favoured in a metal with a strong electron–phonon interaction. Hence the better superconductors are often rather poor normal conductors, such as mercury (transition discovered by Kamerlingh–Onnes just above 4 K), niobium (transition temperature $T_C = 9$ K), tin (3.7 K), lead (7.2 K). The group 1 metals (alkali metals and noble metals) probably never become superconducting, however cold they become.

A feature of the BCS theory (and indeed of our understanding of all superfluids, whether formed from fermion pairs or from bosons) is that the condensate is described by a simple coherent macroscopic wave function. The word "macroscopic" describes the idea that phase coherence of the function is maintained over the whole of the condensate, so describing quantum mechanics in teacups, not just in atoms! A dramatic demonstration of this large-scale coherence is found in what we now call the Josephson effects. These concern the behaviour of two lumps of superconductors (or any superfluid) connected by a "weak link". Brian Josephson was awarded the Nobel Prize for Physics in 1973, the year after BCS's recognition, for "his theoretical predictions of the properties of a supercurrent through a tunnel barrier, in particular those phenomena which are generally known as the Josephson effects."

In liquid ^{3}He, each of the atoms has a nuclear magnetic moment and the liquid is quite strongly magnetic. Hence one atom moving through the liquid polarises the surrounding fluid. This magnetic polarisation lasts transiently after the atom has left the scene, and can be sensed by a second atom arriving later. Thus, pairs of atoms can be formed, even though there is no permanent binding between them, and ^{3}He undergoes a superfluid transition. However, like superconductivity, this is a subtle and small effect. The superfluid transition in ^{4}He is at around 2 K, but in ^{3}He T_C occurs only at 1–2 mK, three orders of magnitude lower. Its discovery in 1971 by David M. Lee, Douglas D. Osheroff and Robert C. Richardson was finally recognised by the 1996 Nobel Prize. Superfluid ^{3}He has some similarities with superconductivity, in that it is a pairing superfluid in a Fermi system, but important differences also. It is a highly pure substance and there is no lattice to complicate matters. The atom is electrically neutral, but has magnetic properties. And even more critically, the pairing takes place between atoms with parallel spins whereas in conventional superconductivity the pair is formed from opposite spin electrons. As we shall see in Chapter 5, these features give a rich behaviour to the superfluid state which makes it an exciting test-bed for our understanding of much basic physics.

Superconductivity is not confined to simple metals and alloys. It has been discovered in many other materials, even in some organic materials. But the most exciting discovery in the last 20 years or so has been the emergence of "high temperature superconductivity". In the 1980s, a whole new class of superconductors appeared. These are layered ceramic oxides rather than metals, with compositions close to a metal-insulator transition. Available transition temperatures T_C to superconductivity almost instantly shot up from around 20 K for the best known alloys to over 90 K for the new ceramics. Great excitement was generated, since superconductivity was suddenly within the range of liquid nitrogen, and new engineering possibilities appeared. The important breakthrough was made by J. Georg Bednorz and K. Alexander Müller, who were rapidly awarded the 1987 Nobel Prize. These materials are not only important for applications, but are also still not theoretically understood. Pairs are certainly involved again, but the nature of the pairing mechanism remains a matter of active controversy.

Finally, it is believed that pairing fermion superfluids play their part in astrophysics. At the enormous densities in a neutron star, for example, one's concept of what is a "low" temperature changes significantly from normal intuition. A typical neutron star has 1.4 solar masses within a radius of only about 15 km, leading to densities roughly 10^{14} times that of our planet. The Fermi energy of these neutrons is so high, around k_B times 10^{12} K, that the interior of the hot star, at only around a million degrees Kelvin, is thought to consist of superfluid neutrons with T_C around 10^9 K. Pulsars are thought to be rotating neutron stars which have strong magnetic fields, so two further Nobel prizes here should complete the list (Anthony Hewish, 1974 and Russell A. Hulse and Joseph H. Taylor Jr, 1993)!

1.5 WHAT MAKES A SUPERFLUID SUPERFLUID?

At the start of Section 1.4, we noted that the existence of a condensate, a single coherent ground state, was essential for superfluidity to exist. But what does the word superfluidity mean, and how is it observed?

Of course, it suggests some extraordinary ("super") properties of fluidity, of flow. This implies not only a coherent ground state, which can be deformed a little to generate a current; it also requires that such a deformed ground state does not immediately dissipate itself by forming excitations of the fluid above the ground state. Often this requirement is met by there being an *energy gap* between the ground state and the excited states of the substance. In other words, the nature of the excitations is important as well as the nature of the ground state.

1.5.1 The Landau critical velocity

This argument was brilliantly clarified by Landau, who explained the requirement for there to be a non-zero critical velocity v_L for flow, in other words for there to be a regime of so-called *superflow* at velocities less than v_L.

Of course there are many experiments which could be considered here, but what is important is the *relative* velocity between the fluid and its surroundings. The simplest case to consider is the interaction between a very heavy object and the fluid. This could imply that the fluid moving at velocity v and the surrounding walls (of effectively infinite mass) are stationary. Or it could imply a stationary fluid with a heavy projectile being dragged through it at velocity v. Landau's argument relates to the second case, since it is easier to visualise.

Consider a projectile of mass M which is much larger than the masses of the bosons which make up the condensate of the substance, injected at velocity v. Whether or not the projectile feels a drag force depends on the availability of suitable excited states in the fluid. For an excitation to be created, and hence for dissipation to occur, the spectrum of excited states must allow the process to conserve energy and momentum.

Suppose that the projectile creates an elementary excitation of energy E and momentum $\mathbf{p}$, in the process of which the projectile velocity changes from $\mathbf{v}$ to $\mathbf{v}'$. The conservation laws require:

$$\frac{1}{2}Mv^2 = \frac{1}{2}Mv'^2 + E$$

and

$$M\mathbf{v} = M\mathbf{v}' + \mathbf{p}.$$

Assuming that the kinetic energy of the large projectile is much bigger than that of the elementary excitation, the energy equation approximates to

$$M(v-v') = \frac{E}{v}$$

and the momentum equation may be written as

$$M(\mathbf{v}-\mathbf{v}') = \mathbf{p}.$$

Now it is a matter of simple geometry that

$$|\mathbf{v}-\mathbf{v}'| \geq (v-v')$$

the equal sign being appropriate when the excitation is ejected in the forward direction, so that $\mathbf{v}$, $\mathbf{v}'$ and $\mathbf{p}$ are all parallel. Hence it follows that $v \geq E/p$, in other words that there is a minimum critical velocity for creating an excitation of

$$v_L = \left(\frac{E}{p}\right)_{min} \tag{1.4}$$

This is the Landau critical velocity.

1.5.2 The importance of the dispersion relation

So, when does a condensate show superfluid behaviour? The answer is that superfluidity occurs when the Landau critical velocity is non-zero. This is because if v_L is zero, then the least flow will cause dissipation.

The question thus centres on the spectrum, i.e. the dispersion relation, for the excitations. As indicated in equation 1.4, we need to find the minimum value of (E/p) for an excitation. The dispersion relation may be displayed as a simple graph of E against p, whereas E/p constant represents a straight line through the origin on these axes. The Landau velocity is thus found as the line of minimum slope which intersects the dispersion relation graph.

First consider Bose systems. We can illustrate the situation very easily by discussing the ideal Bose gas. Here, the excited states are simply states of an independently moving gas molecule. In other words, they have the energy equal to the kinetic energy of a molecule. Hence the dispersion relation is $E = p^2/2m$. As was realised by Landau, this dispersion relation has essentially $v_L = 0$, since $E/p = p/2m$ has its minimum at $p = 0$. Any flow can generate excitations of very low energy and hence dissipate (see Figure 1.6A).

This led Landau to propose that the dispersion relation for the interacting boson liquid ^{4}He must be of an entirely different sort. As he suggested,

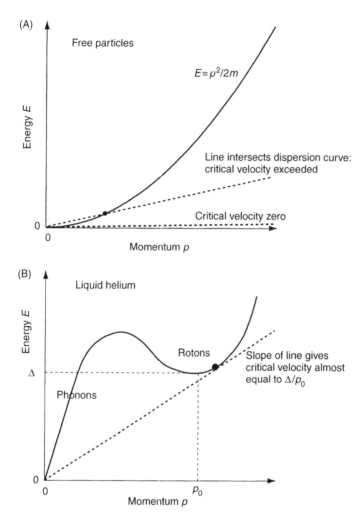

Figure 1.6 Dispersion relations and Landau critical velocities for boson systems. (A) Free particles, showing zero Landau velocity and hence no superfluidity; (B) Liquid helium, showing the phonon/roton dispersion relation with a large critical velocity.

and as we now know from neutron scattering experiments, the excitation spectrum has the form shown in Figure 1.6B. At low energies, the excitations are sound waves, phonons, which approximately obey a linear dispersion relation $E = pc$ where c is the velocity of sound. Note that without other excitations this would give a critical velocity equal to c. But there are also higher energy excitations (somewhat mysteriously called rotons) with a dispersion relation of the form $E = \Delta + (p - p_0)^2/2m$. This leads to a critical velocity approximately equal to $v_L = \Delta/p_0$ as indicated in the figure. We shall see in Chapter 2 that this remarkable picture is robust and accurate. It is a tribute to Landau's genius that he predicted it without any experimental confirmation.

For Fermi systems, a similar idea operates. A central result to the pairing theory of BCS is that there is an energy gap, also written as Δ, between the condensate (the ground state) and the excited states.

The idea goes as follows. The ground state of the Fermi system is no longer a state of zero energy, since the Exclusion Principle demands that the one-particle states are filled up to the Fermi energy E_F. It is often pictured as a Fermi sea, with particles occupying all states with momentum up to the Fermi momentum p_F. For an ideal gas of free particles of mass m, the Fermi sea is a sphere in momentum space of radius p_F where $E_F = p_F^2/2m$. This situation is illustrated in Figure 1.7A. A low energy disturbance of the smooth Fermi sea occurs when a particle below the surface (i.e. with energy just less than E_F) is promoted to an empty state (note the Exclusion Principle constraint here) just above E_F. Thus excitations are generated in hole-particle pairs, with both partners being close to the Fermi surface.

In considering such excitations, it is actually much more convenient to describe matters in an "excitation picture". In this description, all energies are measured relative to E_F, not to the vacuum. Also the picture takes care of the fact that all the states below E_F are already filled in the ground state. The excitation picture in an ideal normal Fermi gas is illustrated in Figure 1.7B. The lowest one-particle excitations are those in the neighbourhood of the Fermi level, now counted as $E = 0$. The pair of excitations discussed above now appears as one on the hole branch (with $p < p_F$) and one on the particle branch (with $p > p_F$). With reference to the Landau criterion, equation 1.4, the diagram makes it clear that we have $v_L = 0$ for the normal Fermi gas. At an arbitrarily low velocity we can generate an excitation pair at the Fermi surface, with energies of each excitation as close to zero as we like, and with momenta essentially equal to p_F.

At last, we come to consider what happens in a Fermi system when the pairing interaction causes the formation of a condensate. The principal result is that the ground state becomes of lower energy than the sum of the kinetic energies of the individual fermions. The small attractive interaction, when all the particles cooperate, gives a modest stability to the condensed state. Excitations again occur in pairs, since excitation demands breaking a Cooper pair from the condensate. But now, this excitation pair no longer has zero threshold energy. Even though both partners can be at momentum p_F, the reduction in numbers in the condensate requires a minimum energy Δ (of order of magnitude $k_B T_C$ as we shall see later)

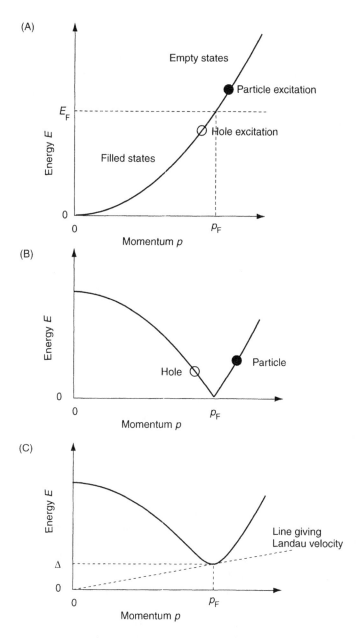

Figure 1.7 Dispersion relations and Landau critical velocities for fermion systems. (A) Free particles in the vacuum picture; (B) Free particles in the excitation picture; (C) The excitation picture when pairing causes an energy gap. There is now a Landau critical velocity, equal to Δ/p_F.

for each excitation. This is illustrated in Figure 1.7C. [For enthusiasts, the expression for the excitation energy becomes $E^2 = \Delta^2 + (|(p^2 - p_F^2)|/2m)^2$.] The Landau criterion now gives a non-zero critical velocity, with the onset of dissipation occurring with the formation of two excitations, each with $E = \Delta$ and $p = p_F$ in the direction of relative motion. Hence we have $v_L = \Delta/p_F$.

1.6 SUMMARY

In this chapter we have introduced some of the features which fascinate low temperature physicists. In brief:

1 At a low enough temperature, everything which remains in thermal equilibrium must enter an ordered state. This relates to the Third Law of Thermodynamics.

2 Fluids exist even at $T = 0$, which raises interesting questions about the meaning of an "ordered fluid".

3 Quantum mechanics (zero-point energy) makes solid helium unstable at normal pressures, so that liquid helium is one such fluid.

4 Actually liquid helium is two such fluids, since quantum mechanics demands that ^{3}He (a fermion) and ^{4}He (a boson) have dramatically different properties.

5 Many fluids order by becoming superfluids.

6 For bosons, superfluidity is a major effect related to the Bose–Einstein condensation.

7 For fermions, superfluidity is possible in a more subtle way by the formation of Cooper pairs (two odds make an even).

8 One hallmark of a superfluid is the description of the superfluid ground state by a single coherent macroscopic wave function.

9 Superfluid flow properties occur if the dispersion relation for excitations above the ground state takes an appropriate form, such that the Landau critical velocity is non-zero. An energy gap in the excitation spectrum can do the trick.

10 Common superfluids include: liquid ^{4}He, cold atoms (bosons) and superconductivity, liquid ^{3}He, neutron stars (paired fermions).

In the rest of this book, we shall explore three of these superfluids in some detail. In Chapter 2 we look at superfluidity in liquid ^{4}He. Chapter 3 is a diversion to introduce the physics of some low temperature techniques, before investigating superfluid fermion systems. Superconductivity is discussed in Chapter 4 and finally liquid ^{3}He in Chapter 5.

FURTHER READING AND STUDY

For useful background, consult any standard text on thermal and statistical physics. Books with a low temperature material include:

J. S. Dugdale, *Entropy and its Physical Meaning,* Taylor and Francis, 1996.
Tony Guénault, *Statistical Physics,* Kluwer, 1995.
C. B. P. Finn, *Thermal Physics,* Nelson Thornes, 1993.
R. H. Dittman and M. W. Zemansky, *Heat and Thermodynamics,* McGraw Hill, 1997 (or earlier editions).
F. Mandl, *Statistical Physics,* Wiley, 1988.

Material for investigating the history and background of the subject includes:

K. Mendelssohn, *The Quest for Absolute Zero,* Taylor and Francis, out of print but an interesting read if available at your library.
On the internet [http://almaz.com/nobel/physics/physics.html] gives a useful resource about Nobel prizes in Physics.
Royal Holloway (University of London) have introductory low temperature material produced under a Partnership for Publication Awareness project. There are useful links and information at [http://www.ph.rhbnc.ac.uk/lowtemp/posters/].

PROBLEMS

Q1.1 The triple point of nitrogen (Figure 1.1) occurs at 63 K and at a pressure of 124 mbar. Use this data to estimate dP/dT and hence the latent heat of evaporation of nitrogen in kJ per liquid litre at atmospheric pressure. The density of the liquid is $790 \, \text{kg m}^{-3}$. Explain why the actual value $(160 \, \text{kJ L}^{-1})$ is considerably larger.

Q1.2 The dispersion curve near the roton minimum in superfluid ^{4}He can be expressed as $E = \Delta + (p - p_0)^2/2m$. At zero pressure, the numerical values are given by $\Delta = 8.7 \, \text{K}$, $p_0/\hbar = 19 \, \text{nm}^{-1}$ and $m = 0.16 m_4$ where m_4 is the mass of a helium atom. (i) Estimate the Landau critical velocity from these figures. (ii) Discuss the error (magnitude and sign) made in assuming that this velocity is simply Δ/p_0.

Q1.3 In the text we stated the dilemma of the Third Law, whether it is (i) true but useless, or (ii) untrue but nevertheless useful. Explore this problem with examples.

Q1.4 Discuss in your own words (i) why helium remains liquid at the lowest temperature and (ii) why the molar volume of liquid ^{3}He is larger than that of liquid ^{4}He, even though the interatomic forces are identical.

Q1.5 Core physics topics to give useful background

   (a) Entropy and disorder
   (b) The laws of thermodynamics
   (c) The Kelvin temperature scale
   (d) Derivation of the Clausius–Clapeyron equation
   (e) Quantum mechanics of identical particles – fermions and bosons
   (f) Fermi–Dirac and Bose–Einstein distributions for ideal gases

   (g) The Heisenberg uncertainty principle
   (h) Dispersion relations

Q1.6 Possible essay topics.

   (a) Phase diagrams and the Clausius–Clapeyron equation
   (b) The Landau critical velocity – a criterion for superfluidity?
   (c) The helium liquids are often called "quantum liquids". Why?

Chapter 2

Liquid ^{4}He

The history of the discovery of superfluidity in liquid ^{4}He makes an extraordinary tale. Helium was first liquefied by Kamerlingh Onnes in Leiden in 1908. At the time this feat was a remarkable achievement in itself, but Onnes continued to cool the newly discovered liquid further, in an attempt to produce solid ^{4}He. To do this he reduced the pressure, expecting solid to be produced as soon as the triple point was reached (compare Figure 1.1). As we now know, he was doomed to frustration in this endeavour, because there is no triple point (Figure 1.2). Nevertheless, he did in fact cool the liquid to around 1 K, well into the superfluid phase, but did not recognise the phenomenon.

Over the next 30 years, this remained the situation. Various clues were found, but not understood and not always published. Principal among these clues were the specific heat maximum at 2.17 K, giving rise to the name λ-point, discovered in Leiden by Keesom around 1930. Even before that Onnes himself had noted a maximum in the density of the liquid. But the spectacular sign should have been the extraordinary ability of the liquid below the λ-point to conduct heat, so marked that boiling occurs not by bubble formation but by steady evaporation from a placid surface. In spite of this clue, the nature of ^{4}He below the λ-point (so-called He II) remained unexplained.

In the event it was not until 1938 that the nature of He II was recognised as a manifestation of superfluidity. This arose from measurements of viscosity. Two papers were published side by side in *Nature* that year, one by Allen and Misener (Cambridge) and the other by Kapitza (Moscow), who coined the word "superfluidity" to describe the observed properties. Shortly afterwards Fritz London worked out the theory of an ideal Bose–Einstein gas, explaining the phenomenon of Bose–Einstein condensation. Tisza carried the argument further by suggesting a "two-fluid model" for liquid He II, a suggestion made independently (communication was at best slow at this time between East and West) and worked out in considerable detail by Landau in Moscow.

Why was there this 30 year gap following the liquefaction of ^{4}He? It is fascinating to speculate with the hindsight of history. I suppose that the discovery by Kamerlingh Onnes in 1911 of superconductivity produced enough excitement to occupy the mind. It is also important to realise what a small community was involved in these matters, with only a handful of laboratories performing these studies (notably Leiden, Toronto, Cambridge, Oxford, Moscow)!

In this chapter we shall first outline briefly some basic experimental properties of liquid ^{4}He and their description by the two-fluid model. Next we discuss the behaviour of excitations in liquid helium and the breakdown of superfluidity. Finally we describe the idea of pure superfluid helium as a quantum fluid and the quantisation of vorticity.

2.1 SOME PROPERTIES OF LIQUID ^{4}He IN THE TWO-FLUID REGION

The idea of a two-fluid model was suggested by the properties of the Bose–Einstein condensation of an ideal gas. Below the transition temperature, the gas consists of (1) a substantial fraction of its particles in the ground state and (2) the remainder occupying excited states. The ground state fraction f varies with temperature from zero at the transition to unity at the absolute zero. For a Bose–Einstein ideal gas, one finds $f = 1 - (T/T_0)^{3/2}$ where T_0 is the transition temperature. Thus the thermal properties can be described as a sum of the ground state fraction which is ordered, and hence makes zero entropy contribution, together with the "normal" fluid fraction, consisting of particles in the excited states, which has a conventional entropy.

What Tisza and Landau appreciated was that many of the peculiar properties observed in liquid He II can start to make sense in the context of a similar two-fluid model. In this model one considers the helium to consist of two interpenetrating fluids, called the "normal" fluid and the "superfluid".[1] The two fluids are postulated to have the properties given in Table 2.1.

We shall now see how these postulates may be used to understand the observed properties of He II.

---

1 One should note here a certain ambiguity in use of the word "superfluid" in this context. It is used sometimes to mean only the superfluid fraction, but at other times to mean the whole fluid when it is below its transition temperature. In this section we shall use it in the first sense only, and refer to the whole fluid as He II. Otherwise, the context must decide.

Table 2.1 The two-fluid model

	Normal fluid	Superfluid	Relationship
Density	ρ_N	ρ_S	$\rho_S + \rho_N = \rho$
Velocity	$\mathbf{v}_N$	$\mathbf{v}_S$	$\rho_S \mathbf{v}_S + \rho_N \mathbf{v}_N = \rho \, \mathbf{v}$
Viscosity	η_N	$\eta_S = 0$	
Entropy	S_N	$S_S = 0$	$\rho S = \rho_N S_N$

2.1.1 Mechanical properties

One of the early indications of the peculiarity of He II was found in the attempted measurement of viscosity. The coefficient of viscosity η is defined as the ratio between the shear stress on a fluid to the rate of change of the shear strain (i.e. the fluid flow). Low viscosity means ready flow, high viscosity means reluctant flow. One commonly measures η in one of the two ways, as illustrated in Figure 2.1.

1. Flow through tubes The classic method is to establish a pressure difference between the ends of a tube and to measure the rate of flow of fluid through the tube. The viscosity is essentially a measure of the flow resistance, i.e. the ratio of driving force to observed flow. For a normal ("Newtonian") fluid, such as liquid helium above the λ-point, the coefficient so defined is independent of the pressure difference, i.e. the flow is proportional to the force. What happens in He II is quite different. The flow is very much larger than that expected from measurements above T_λ. Furthermore, it is not proportional to applied pressure, rather the flow becomes large at the smallest applied pressure difference and then saturates, staying effectively constant when further pressure is applied.

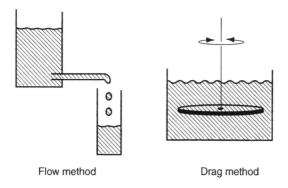

Flow method   Drag method

Figure 2.1 Methods for measuring viscosity.

The flow is clearly limited by other effects (see Section 2.3.6). In the helium cryostat trade, this is an example of a "superleak", since He II is found (depressingly) to penetrate even the smallest leak in an apparatus. He II behaves indeed as a "superfluid" in this context.

2. Damping of a vane   The other well-known method to determine the viscosity of a fluid is to measure the drag force on a moving object. The cleanest experiment is to observe a torsional oscillator, usually a flat circular disc suspended by an axial fibre. The disc is thus free to oscillate without displacing any fluid. However, the fluid in contact with the disc must move with the disc, causing a shear strain rate in the nearby fluid. The associated viscous shear forces resist any movement of the disc and thus cause the oscillation to decay. Hence the damping of the disc is related directly to the viscosity of the fluid. When this experiment is performed in helium *above* the λ-point, it yields (of course) precisely the same result as method 1 above. However, as the temperature is reduced through T_λ there is no sudden change in damping and thus in measured viscosity. This result is in dramatic contrast to method 1. We may note that as the temperature is lowered further, there is a small and gradual decrease in damping, but nothing sudden or dramatic.

These observations are precisely those to be expected from a two-fluid model. In any mixture of two fluids, method 1 must be dominated by the smallest of the two viscosities, since the thinner fluid can find its way through the tube much more readily than the thicker one. On the other hand, method 2 will be dominated by the largest of the two viscosities, since the large drag force exerted by the thicker fluid will prevail over the smaller force from the thinner one. In an electrical circuit analogy, method 1 represents two resistors in parallel, and the current is dominated by the channel of least resistance. Method 2 is like having two resistors in series, since the disc must plough its way through both fluids, and the larger resistance will dominate. Of course, the two fluids in helium are not two chemically distinguishable fluids. The situation is more subtle than that, since all the helium atoms are identical and the two fluids cannot even in principle be separated. Nevertheless, the two experiments are well described using the properties in Table 2.1, with the superfluid component having zero viscosity. The additional ingredient is that there must be a mechanism which limits the flow in method 1. This mechanism implies a breakdown in superfluidity. In practice in this type of measurement the breakdown occurs by the creation of a tangle of vortices in the fluid when the flow velocity becomes large enough.

3. Andronikashvili's experiment   Following the suggestion of the two-fluid model, a beautiful experiment was carried out in 1946 by Andronikashvili which demonstrated the validity of the model. This uses a set-up similar to method 2 for viscosity, but instead of using a single disc as the oscillating object, a whole stack of discs was used. The experiment is illustrated schematically in Figure 2.2.

The measurement to be made is now not the damping of the oscillator, but its frequency as a function of temperature T. For a torsional oscillator, the periodic time is simply equal to $2\pi\sqrt{(I/k)}$ where I is the moment of inertia of the oscillator and k is the torsion constant of the suspending fibre. The torsion constant does not vary in the experiment, so that the frequency directly yields the moment of inertia, and hence the mass, of whatever is oscillating. The stack of discs in the experiment are made sufficiently close together that the normal fluid is "clamped" to the discs; it is constrained to move with the discs by the viscous forces. As an aside we may note that the condition for clamping, well understood from classical physics, is that the "viscous penetration depth" $(=\sqrt{(2\eta/\rho\omega)}$ where ω is the angular frequency) should be much greater than the disc separation. In Andronikashvili's original experiment, the torsional oscillator had a period of around 30 s, corresponding to a penetration depth of almost a millimetre, whereas his discs were more closely spaced, about 0.2 mm apart.

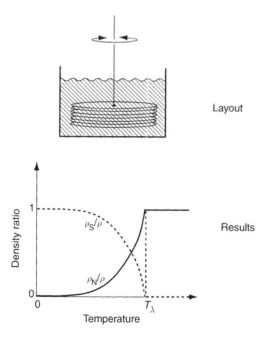

Layout

Results

Figure 2.2 Andronikashvili's experiment.

Above the λ-point, the frequency of such an oscillator is found to be nearly independent of temperature, corresponding to the oscillator having a mass equal to its bare mass plus the mass of the entrained helium liquid. However, below the λ-point, the frequency increases rapidly, indicating the fall in normal fluid density ρ_N. The point is that although the normal fluid is clamped to the discs, the superfluid is not. Instead it remains stationary in the surrounding vessel and thus does not contribute to the effective moment of inertia of the oscillator. As a result, the varying frequency with temperature can easily be converted to a measure of ρ_N versus T. Typical results are shown in Figure 2.2. This experiment convincingly verifies the idea of the two-fluid model. It also enables ρ_N, and hence by subtraction ρ_S, to be determined quantitatively.

4. *"There's a hole in my bucket?" – Film flow*   A remarkable consequence of the ability of the superfluid to flow through very fine channels, such as a superleak, is seen in the phenomenon of film flow. Many liquids wet the surface of their container, and when they do so a thin surface film covers the surface, held there by surface tension forces. Helium is no exception. Above T_λ this has no interesting consequences, however, since the film is sufficiently thin that it is static, clamped to the containing vessel. However, below T_λ the situation is dramatically different. Gravitational syphon action can allow significant flow through the film. For instance a bucket of He II suspended above the free liquid surface gradually empties itself, even without there being any hole in the bucket. [Imagine your surprise if coffee sitting in your mug of coffee were to find its way similarly to the table top.] The amount of flow depends markedly on the geometry of the system under consideration. For instance a bucket with a knife-edge rim at the top would lose much less helium than one with a rounded rim, simply because the film will be thinner over the highly curved surface of the knife-edge. The flow is in any case slow, since at a few cm above the liquid surface the film is typically only 10–20 nm thick and the superfluid can only flow freely at speeds up to around $10\,\mathrm{mm\,s^{-1}}$.

2.1.2   Thermal properties

We have noted in Chapter 1 that the specific heat capacity has the large λ-anomaly at 2.17 K. As remarked earlier, the anomaly is more violent than that in an ideal Bose–Einstein condensation. The heat capacity C can be used to find the entropy S simply from the relation $S = \int_0 C/T\,dT$. The infinite heat capacity at the λ-point thus leads to a vertical tangent in the entropy–temperature relation, as sketched in Figure 2.3.

In relating these observations to the two-fluid model, it is worth making two points.

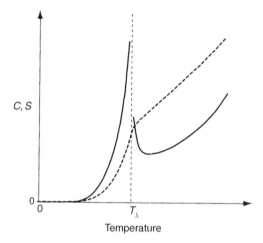

Figure 2.3 Specific heat and entropy of liquid ^{4}He. Heat capacity (full curve) and Entropy (dashed curve).

1   The idea of a zero entropy superfluid, as suggested in Table 2.1, is an appealing one. It certainly gives a valuable way of thinking about the superfluid component. The existence of this superfluid fraction is seen as representing the absolute zero of temperature "coming early". Any fluid must have zero entropy at $T=0$, but here we have this state of complete order intruding at a higher temperature.

2   Nevertheless, the simple identification of $S = \rho_N S_N / \rho$ leads to interesting questions when we apply it to helium, as opposed to the ideal Bose gas. If we accept the clean determination of ρ_N by Andronikashvili's experiment, then we can deduce the properties of S_N. It is particularly interesting to look at the region just above T_λ. In this region, the Andronikashvili experiment is flat; it simply measures $\rho_N = \rho$, constant. However, the evidence from the heat capacity and entropy is otherwise. As the temperature is lowered to just above T_λ, it is as if the system "sees the transition coming" in helium, whereas there is no such precursor in the ideal gas picture. [This is reminiscent of the transition to ferromagnetism in, say, iron; the observed thermal properties around the Curie temperature similarly display a λ-anomaly, whereas a simple mean field theory gives a sudden onset.] This behaviour arises from the fact that helium atoms in the liquid are interacting with each other, so that independent one-particle thinking is bound to be an oversimplification. We may note here that

similar precursors are found in many other properties (e.g. viscosity, as measured by either method; velocity of sound).

This section then contains a gentle warning not to take the two-fluid model too literally. Nevertheless it will remain very valuable in thinking about heat transport and other effects which we discuss in the following sections. We should remark that heat transport in liquid He II is not included in this section as an adjunct to heat capacity. Heat flow in He II turns out not to bear any relation to conventional thermal conduction in a normal fluid, but a new mechanism operates which is a specifically two-fluid effect.

2.1.3 Thermo-mechanical effects

So what happens when heat is applied, say through an electrical heater, to liquid He II? The answer from the two-fluid model is devastatingly simple. The heater merely converts superfluid to normal fluid. Zero entropy superfluid is converted to the entropy-carrying normal fluid at a rate sufficient to absorb the applied energy. Thus an excess of normal fluid and a deficiency of superfluid exists near the heater. As a result a *counterflow* of normal and superfluid is set up as illustrated in Figure 2.4. Superfluid is drawn towards the heater, converted into normal fluid, thus causing a flow of normal fluid away from the heater. In a simple situation where there is no net flow of helium, then the velocities of the two fluids satisfy the expression $\rho_S v_S + \rho_N v_N = 0$. A useful way of looking at this is to realise that the superfluid rushes in to dilute the normal fluid near the heater, since "hot" means low ρ_S and "cold" means high ρ_S, as shown by the curves of Figure 2.2 derived from the Andronikashvili experiment.

   This straightforward idea turns out to have dramatic consequences for the properties of He II, some of which follow.

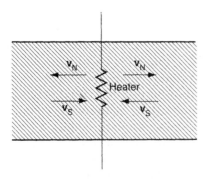

Figure 2.4 Heat transport by superfluid counterflow.

1. Heat transport   The first property, already mentioned, is the abnormally efficient ability of He II to transfer heat. The explanation follows directly from Figure 2.4. The almost frictionless counterflow of normal and superfluid in the neighbourhood of a heater represents a very efficient transport of thermal energy in the liquid. So it is that a local hot spot in the liquid simply cannot occur; the counterflow mechanism irons out any temperature inhomogeneity. Hence the very striking observation that bubbles (which indicate local hot spots in a boiling liquid) are never seen in boiling He II. Instead there is a rapid but smooth evaporation from the surface.

Of course, the heat transfer is never infinite. The counterflow is limited. This is because friction between the two fluids appears when a certain critical value of the relative velocity $v_S - v_N$ is reached. In practice this velocity is a few millimetres per second, and represents the velocity at which self-sustaining tangles of vortex lines are generated in the experimental geometry.

2. The fountain effect   The heat transport property has demonstrated that a temperature gradient provides a driving force for superfluid to be drawn towards the high temperature end of a vessel. This can be demonstrated and measured directly as follows. Consider the experiment shown schematically in Figure 2.5A. This illustrates two helium reservoirs connected by a superleak. The superleak consists of a fine powder packed tightly in a tube, or some other restriction, which is able to allow the passage of superfluid but which clamps the normal fluid. When heat is applied to one reservoir, then the mechanism already mentioned means that superfluid is converted to normal fluid in that reservoir. Thus there is a concentration gradient of superfluid across the superleak, and the superfluid flows through it into the heated reservoir. In the classic fountain effect experiment, illustrated in Figure 2.5B, the consequent inflow of helium causes a fountain of helium to be ejected continuously from the narrow top of the vessel.

Suppose that, in the controlled experiment of Figure 2.5A, the temperature is maintained in the two reservoirs. The flow through the superleak now takes place until a steady state is reached, in which the pressure increase in the heated vessel (the so-called fountain pressure) is just enough to stop the flow through the superleak. A temperature difference is now neatly balanced by a pressure difference, a "thermo-mechanical" effect. We discuss the magnitude of these effects in Section 2.4 below.

3. The inverse fountain effect   In the fountain effect, a temperature difference is seen to generate a pressure difference. But the inverse is also true. Without heat applied, the two reservoirs in Figure 2.5A have equal levels and of course equal temperatures. Now if we apply a pressure difference to one side, we know that only superfluid will pass through the superleak.

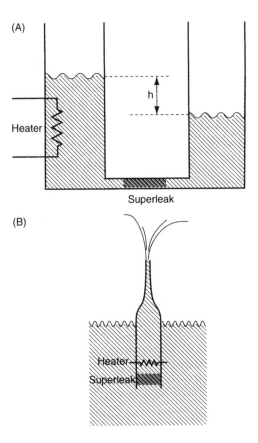

Figure 2.5 The fountain effect. (A) The principle; (B) A demonstration experiment.

Hence we will deplete the superfluid on the high pressure side, i.e. we will increase its temperature. Correspondingly we obtain an excess of superfluid on the low pressure side, so that the liquid there cools. We thus have a simple mechanically operated reversible heat pump. A pressure difference generates a temperature difference.

It is amusing to consider the consequences of this further. If we allow a vessel of He II to empty by gravity through a superleak, then the leaking helium must be pure superfluid. Hence, in an ideal world, it would be at absolute zero. At a first glance, this sounds like a good method of refrigeration. However it actually isn't really very useful since: (1) the superfluid has zero entropy and hence has no useful cooling power to get anything else cold, and (2) we do not live in an ideal world anyway, meaning that in practice some normal fluid is also formed, by excessive flow velocities as the liquid leaves the superleak.

4. *Second sound*   The almost frictionless counterflow of normal and superfluid results in a new type of propagating wave being possible in He II. This counterflow wave is called "second sound", and it is a propagating entropy–temperature wave. This contrasts with normal sound, often called "first sound" in this context, which is a pressure–density wave, and in which the normal and superfluid move together.

An arrangement to demonstrate and measure second sound is sketched in Figure 2.6A. It shows a closed cylindrical tube, with a heater at one end and a thermometer at the other. We have already noted that heat causes a counterflow of normal and superfluid components of He II. When an AC current of frequency f is applied to the electrical heater, the conventional Joule heating has a strong component at frequency $2f$. This thus generates a counterflow, which also has a frequency component at $2f$. The tube can act as a resonator, just like an organ pipe does for normal sound. The standing waves in this case represent regions of oscillation in the fraction ρ_S/ρ_N whilst keeping the total fluid density constant. In other words, a standing temperature wave is set up, which

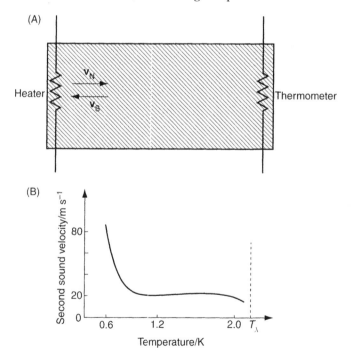

Figure 2.6 Second sound in He II – a counterflow wave (A) Schematic set-up; periodic (AC) heating gives a periodic superfluid counterflow, setting up standing waves detected on the thermometer when the AC has the right frequency; Hence the velocity of second sound is deduced, with the results shown in Figure (B).

can be detected as a $2f$ signal on the thermometer. A simple experiment can demonstrate the existence of these resonant modes in the tube, and hence the velocity of the second sound can readily be determined. Results are illustrated in Figure 2.6B. Second sound is seen to have a velocity of about $20 \, \text{m s}^{-1}$, a factor of about 10 less than first sound, which varies between 220 and $240 \, \text{m s}^{-1}$ between T_λ and zero temperature.

Second sound is strong where there is plentiful normal and superfluid. Thus at high temperatures, just below the λ-point where there is little superfluid, the mode is strongly damped and the velocity becomes even smaller. On the other hand at low temperatures there is little normal fluid to move, so little counterflow is possible and the mode merges with first sound.

Actually the resonator tube experiment can be used to observe both first and second sound with the same equipment. In the set-up of Figure 2.6A we can replace the source and detector by an oscillating superleak. This consists of a metallised piece of a very fine filter (called nucleopore) which acts as a capacitor plate. The nucleopore can be driven or detected electrostatically by application of suitable voltages to the capacitor. As it moves, it allows superfluid to pass through its fine pores whilst the normal fluid cannot do so; thus a combination of counterflow motion (second sound) and bulk flow (first sound) can be excited.

5. *Other sound modes* The existence of the two fluids which can move independently opens up the possibility of yet more propagating modes in addition to first and second sound. Unsurprisingly two more are called third and fourth sound. In these the normal fluid is fully clamped.

Third sound occurs in the helium film on a surface. The film is sufficiently thin that the normal fluid is clamped to it. However, application of a temperature gradient causes superfluid to flow towards the hot part of the film. This can occur (with virtually no pressure difference) simply by the film becoming thicker where it is heated and thinner where it is cooled. Thus a thickness–temperature wave can be sustained in the film, and this is called third sound.

Fourth sound is a bulk effect in a superleak material. Again the normal fluid is clamped and only the superfluid can move. The normal fluid maintains a constant density throughout. But application of a temperature gradient again causes flow of superfluid towards the hotter part. This now must produce a density increase in the hotter part, since the normal fluid cannot flow away as it does in second sound. Hence we now have a propagating density–temperature wave, and this is called fourth sound.

2.1.4 So what is wrong with the two-fluid model?

The idea of the two-fluid model works very well conceptually for the ideal Bose–Einstein gas. And we have seen that it is a useful picture to understand many properties of liquid He II. However, as already remarked, it must not be taken too literally. This is because liquid helium is a strongly interacting system, not one made up of ideal gas atoms. The interactions have two important effects.

One effect, already mentioned in our discussion of the thermal properties in Section 2.1.2, is that above T_λ and close to it, the properties of the liquid are not entirely "normal". As in other interacting systems, the cooperative behaviour of the helium near the transition is enhanced by the presence of short-range interactions, giving the much more violent λ-transition than expected from either long-range interactions (second-order transition as in superconductivity) or from ideal gas statistics (the third-order Bose–Einstein condensation). On cooling towards these more gentle transitions, there is no indication that a transition is close until it actually happens; whereas the heat capacity above a λ-transition diverges above T_λ as well as below.

The second effect is more subtle. It concerns the nature of the super-fluid fraction. Above our picture has been of "all atoms in the ground state", i.e. a condensate which contains 100% of the ^{4}He atoms by the time $T=0$ is reached. Certainly, as we have seen this pure superfluid is thermally dead, well described by $\rho_N = 0$. However, although it is hard to visualise, it seems that interactions have a big effect here. Two types of experiments (neutron scattering and surface tension) can be used to derive a "condensate fraction", that is the fraction of helium atoms which actually occupy the zero-momentum ground state. The surprising answer is that, although the temperature dependence of this condensate fraction mirrors that of the superfluid density ρ_S, the magnitude at low temperatures is not 100% but more like 10%. It is as if the other 90% are dressing the favoured 10% with their interactions. Clearly the one-particle states of the ideal gas model do not tell the whole story. Fortunately, this large discrepancy has no direct effect on the properties or ideas discussed in this chapter, but it does perhaps excuse all the publicity claims in the 1990s about cold atoms giving the first demonstration of a true Bose–Einstein condensation!

2.2 ELEMENTARY EXCITATIONS AND THE CRITICAL VELOCITY

So far we have simply visualised the normal fluid in He II as "a bit like a normal fluid". This is not such a bad picture, as we have seen, to understand the phenomena observed close to T_λ. However it does lack a certain sharpness!

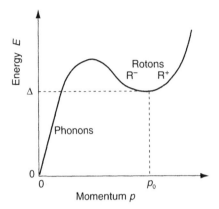

Figure 2.7 The dispersion relation for excitations in liquid ^{4}He.

At low temperatures in particular, in order to model, say, the heat capacity, it is essential to know what the dispersion relation is for the elementary excitations in the helium. In other words, what is the normal fluid.

As already stated in Chapter 1, the answer here is a surprising one. The dispersion curve takes the form illustrated in Figure 2.7 (repeated from Figure 1.6), with a phonon region and a roton region. The phonon region indicates that the long wavelength (small momentum) excitations are the expected first sound phonons, satisfying $E = c_1 p$ with the sound velocity $c_1 \sim 240 \, \text{m s}^{-1}$. But the roton region is unique to helium. We have already seen that this form, predicted by Landau, gives rise to superfluidity with a high critical velocity for breakdown into rotons.

2.2.1 How is the dispersion curve measured?

A direct determination of the dispersion curve in many substances can be made by neutron scattering experiments. The idea is simple enough, although the carrying out of the procedure requires clever equipment to be operated with skill and much patience. The basic scheme is that an incident neutron, of known energy and momentum, is allowed to scatter from the liquid helium. The scattered neutrons, again of known energy and momentum, are observed. Conservation laws then tell us the values of the energy and momentum of whatever excitation caused the scattering. This is a powerful experiment, although a difficult one, since the neutron's energies and momenta before and after the event must be determined with some accuracy. In addition, any scattering which arises from processes other than the creation of elementary excitations in the liquid must be subtracted out. Nevertheless, these experiments have been performed, and the form of Landau's suggestion (Figure 2.7) has been amply justified, and the parameters measured in this direct way.

There is an added bonus in the neutron scattering experiments. The spread of the scattered neutron beam which arises from a particular region of the helium dispersion relation can also be measured, in addition to the mean position of the beam. From the spread one can also determine the lifetime of the excitations. Basically one finds that the excitations are well-defined in the phonon region and around the roton minimum; the higher energy excitations are more short-lived. Furthermore, the basic curve is more or less independent of temperature (even above T_λ, somewhat surprisingly!) although as expected the spectrum becomes broader as temperature rises and lifetimes shorten. Finally we may note that the detail of the dispersion curve does of course depend on density. The values of the initial slope (sound velocity) and the parameters Δ and p_0 of the roton minimum vary somewhat with molar volume. Hence there is also a little temperature variation when measurements are made at saturation vapour pressure.

2.2.2   What properties are understood in terms of the dispersion curve?

As in most substances, a definitive knowledge of the dispersion relation allows us to understand a wide variety of measurements. For example, in a metal or semiconductor, knowledge of the band structure (the dispersion relation for electrons) leads to a compact understanding of many different properties. So here, the dispersion relation is a key feature. We have already seen that the somewhat bizarre shape explains why helium becomes a superfluid with a non-zero critical velocity, the Landau velocity. But it allows us to have a unified understanding of many other properties, some of which are mentioned below.

1. Thermal properties   The occupation of states described by the dispersion relation can be described by simple statistical physics. This is a valid approach at low temperatures, where the elementary excitations are well-defined and can be thought of as independent entities (quasiparticles) in thermal equilibrium to which statistical methods therefore apply. In a way, this can be thought of as a way of calculating the "normal fluid" density ρ_N at low temperatures.

This now means that the thermodynamic properties (entropy, heat capacity) may be calculated from the dispersion relation. The linear portion of the curve, corresponding to phonons, yields a T^3 heat capacity contribution which dominates at low enough temperatures, in practice below about 0.5 K. This is the usual Debye approximation, often applied to a crystalline solid, but here even more reasonably applied to a continuum. Above this temperature, rotons start to play a part, dominated by a term $\exp(-\Delta/k_B T)$ where Δ is the roton minimum energy. The roton contribution grows to equal the phonon contribution at around 1 K. This whole treatment is found to be in good agreement with experiment, a triumph for the unusual dispersion curve. This simple treatment works well at temperatures up to about 1.3 K, above which the more collective philosophy of the two-fluid model becomes more appropriate.

In passing we may note that many transport properties (e.g. viscosity and thermal conductivity) can also be successfully described at low temperatures in terms of the kinetic behaviour of a gas of these elementary excitations.

2. Quantum evaporation   A further way of exploring the dispersion curve has emerged in recent years with some beautiful measurements on what is called "quantum evaporation". Although there is much of interest also in the detail of these experiments, we would just discuss the idea here.

A bath of He II is prepared and held at low temperature (typically below 100 mK) with a free and stable surface between liquid and gas. Actually the free surface is effectively between pure superfluid helium and vacuum, because the saturation vapour pressure is so low. A collimated heat pulse is applied within the liquid. Essentially this is a stream of ballistic excitations, because there are so few pre-existing thermal excitations at the low temperature. What happens when these excitations reach the free surface? The answer for many of them is quantum evaporation. This is a process in which an incoming excitation in the liquid is absorbed by a single helium atom. The atom then can have enough energy to be ejected from the surface into the vacuum, where its arrival at a detector can be observed. Only high energy excitations are candidates for this process since the binding energy of the atom (around 7.1 K) must be exceeded for evaporation to take place.

There are two particularly clever features of the experiment. One is that, by using heat pulses, time-resolved measurements can be made, allowing a distinction between high energy phonons and rotons. The velocity of the excitation is the group velocity dE/dp, the slope of the dispersion curve, quite different for the two excitation types. Varying the position of the helium surface gives a further diagnostic here. The second feature is that by careful arrangement of the source/surface/detector geometry, one can determine the angle of "refraction" at the surface, i.e. the angle which the evaporated atom makes to the surface, compared to the angle of incidence of the excitation.

There are several subtleties here, all of which derive from simple principles of conservation of energy and momentum. One such concerns the excitation beam itself, which contains large numbers of high energy

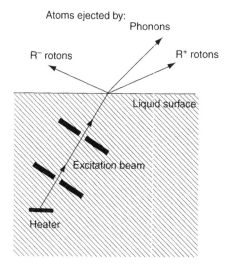

Figure 2.8 A quantum evaporation experiment.

phonons. This is because there is a slight upward curve (unusually) in the phonon dispersion curve, which means that simple two-phonon decay processes, which split one high energy into two of lower energy, are not possible; only much less probable three-phonon processes will do the trick.

The second concerns the so-called R⁻ rotons, excitations on the low momentum side of the roton minimum. These rotons have a negative group velocity, i.e. their velocity of propagation is opposite to their momentum. When conservation of transverse momentum at the helium surface is invoked, it is clear therefore that evaporated atoms must be emitted backwards from the surface as illustrated in Figure 2.8. These features, reinforced by the time of flight measurements, are entirely consistent with the form and detail of the dispersion curve.

3. Ion mobility and the Landau critical velocity   Another direct measurement of significance is the study of "ions" in liquid helium. In this type of experiment, ions are injected into the liquid by applying a high voltage to a fine metal tip. The ion motion to a collector electrode can be controlled and studied with suitably arranged grids set at specific potentials. The idea is sketched in Figure 2.9.

Now the "ions" which travel in the liquid are not simply atomic helium ions (helium atoms with one too many or one too few electrons). Instead they turn out to be very heavy objects. The positive ions start off as a helium atom with one electron stripped off it, but the electric field around this atom is enough to attract the surrounding liquid so strongly that a solid

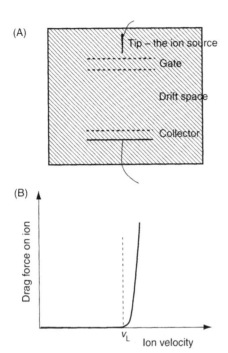

Figure 2.9 An ionic mobility experiment: (A) Schematic set-up; (B) Results showing the Landau velocity.

"snowball" of over 50 atoms is formed. On the other hand a negative ion starts off by being a free electron, but an electron can only be localised if it has a high momentum (Heisenberg's Uncertainty Principle). Thus the electron is so active that it creates a large bubble, a hole, in the helium which excludes around 100 helium atoms. Therefore it turns out that both positive and negative ions, although small enough to our eyes, are objects which have a very large effective mass compared to one helium atom. The precise size of these ions depends on pressure applied to the helium.

In the experiment, the ionic mobility is measured. What processes can slow down one of these massive ions? At low enough temperatures in He II, in the absence of ^{3}He impurities, and (as it turns out) at pressures above about 21 bar, the loss mechanism for negative ions is the generation of elementary excitations in the helium. As discussed earlier, this is a process which must conserve energy and momentum, so that it is subject to the constraints of the dispersion relation. Furthermore, the interaction between an excitation and a large object is precisely the situation discussed by Landau, as outlined above in Section 1.5.1. Because of the nature of the dispersion curve, no such processes are possible until the Landau velocity ($v_L = \Delta/p_0$) is reached. Note that v_L has the heroic value of around $50\,\mathrm{m\,s^{-1}}$ in liquid ^{4}He, the same sort of speed as a car on the motorway. The measurements of negative ion mobility provide a beautiful verification of this value. There is virtually no dissipation seen until the Landau velocity is reached, above which value it cuts in as illustrated in Figure 2.9B. This is a very pretty direct measurement of the Landau velocity, and the magnitude agrees remarkably well with that calculated from the dispersion relation.

2.3 QUANTUM EFFECTS, VORTICITY AND ROTATION

So far in this chapter, we have taken the existence of a superfluid component in ^{4}He as an experimental fact, as evidenced by the two-fluid model. But, as stressed in Chapter 1, the "explanation" for the existence of the superfluid is rooted deeply in quantum behaviour.

In a single phrase, the postulate here is that the whole bath of superfluid ^{4}He can be described by a single coherent wave function, with the simplest form:

$$\Psi = C \exp(i\phi) \tag{2.1}$$

In this equation both the amplitude C and the phase ϕ can be functions of space and time. The amplitude relates to the strength of the superfluid, i.e. $C^2 = |\Psi|^2$ represents the superfluid (number) density, which may be written in terms of the superfluid density ρ_S as $C^2 = \rho_S/m_4$ where m_4 is the mass of a ^{4}He atom. But the amazing feature is the existence of a well-defined phase function ϕ. This implies macroscopic quantum behaviour, in other words that the fluid at one side of a container "knows about" the properties of the superfluid at the other side, a macroscopic distance away. This is the essence of superfluidity.

To understand more, we need to recall the physical significance of the phase of a wave function. Roughly speaking, it is valid that wiggles (spatial contortions) in a wave function represent momentum. So for a free particle, the spatial variation of the wave function is $C\exp(i\mathbf{k}.\mathbf{r})$, or $C\exp(i\mathbf{p}.\mathbf{r}/\hbar)$, where $\mathbf{k}$ is the wave vector and $\mathbf{p}$ is the momentum of the particle. The higher the momentum of the particle, the more rapidly does the wave function oscillate in space. Hence it is to be expected that the phase ϕ should relate to the momentum of the superfluid, and hence to the superfluid velocity $\mathbf{v}_S$.

The obvious, and as it turns out, the correct identification is to write the momentum $\mathbf{p}$ simply as $m_4\,\mathbf{v}_S$. This identification then leads to

$$\phi = \frac{\mathbf{p}.\mathbf{r}}{\hbar} = \left(\frac{m_4}{\hbar}\right)\mathbf{v}_S.\mathbf{r}. \tag{2.2}$$

To find $\mathbf{v}_S$ we invert this equation using the gradient operator to give

$$\mathbf{v}_S = \left(\frac{\hbar}{m_4}\right)\nabla \varphi \tag{2.3}$$

Hence we see that the superfluid velocity is directly related to the gradient of the superfluid wave function.

As an aside here, let us remark that this expression can be derived much more rigorously using the appropriate quantum mechanical current operator. It is then clear that we have neglected the influence of any rapid change in Ψ arising from a change in C or equivalently ρ_S, but this neglect is valid for most situations. We have also tacitly assumed that m_4 is the correct mass to go into the equation, another assumption that is appealing and correct, but not immediately obvious.

2.3.1   Zero circulation?

The identification of phase gradient with superfluid velocity has an immediate consequence. This concerns the "circulation" of the fluid. Circulation is defined as the loop integral $\oint \mathbf{v}_S.d\ell$ where $d\ell$ is a vector element along the length of the path. From equation 2.2 we can see that the circulation simply takes the value of $(\hbar/m_4)$ times the phase difference around the loop. Now consider loops entirely within a simply connected region of the fluid (see Figure 2.10A). For the wave function of equation 2.1 to make physical sense, the phase difference around a loop must be zero or an integral number of 2π – wave functions around loops must eat their tails to be single valued and thus physical. But within the simple region, with non-zero superfluid density and no holes or singularities anywhere, the only possible result is zero. This is because the loop can be placed anywhere and indeed shrunk to a vanishingly small size. It is a topological necessity.

This last result can be stated very neatly in differential form as

$$\text{curl } \mathbf{v}_S = 0 \text{ or } \nabla \times \mathbf{v}_S = 0. \tag{2.4}$$

This is an intriguing result and one which is often taken as a starting point for working out the behaviour of the superfluid. A fluid whose velocity field is described by equation 2.4 is said to be "rotationless" or to exhibit "pure potential flow". One curiosity is to ask how such a curl-free fluid can rotate. Indeed, without any further ideas it cannot do so. When a bucket of superfluid helium is rotated, the prediction would be for the fluid to stay still (with respect to the fixed stars?) whilst the bucket alone

(A)

(B)

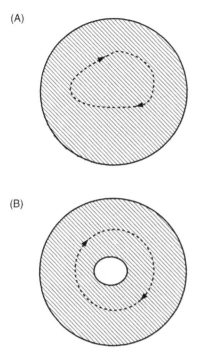

Figure 2.10 Wave functions must eat their tails when a closed circulation path is taken. (A) Simply-connected superfluid; (B) Multiply-connected superfluid.

does the rotation. However, this is not the complete story as we shall now understand.

2.3.2 Quantisation of circulation

Instead of the simple circuit in a simply-connected bucket of superfluid helium, consider now what happens in an annular ring of the fluid. This is now a multiply-connected situation. We can now discuss the circulation of the velocity around paths which go around the "hole" in the liquid, as illustrated in Figure 2.10B. The big difference now is that these paths cannot be shrunk down to nothing; a deformation of the path always leaves it going around the hole.

Hence, when we consider the necessary restriction on the phase, we can again say that the wave function around the hole must eat its tail. In other words, the phase change around the loop must again be an exact integral multiple of 2π. However there is now no requirement that this multiple, n say, is zero. It is a geometrical consequence of the existence of a coherent wave function, as in equation 2.1, which immediately leads us to see that circulation must be quantised. Working out the circulation $\oint \mathbf{v}_S.d\ell$ from equation 2.2, we now obtain

$$\oint \mathbf{v}_S.d\ell = 2\pi n \frac{\hbar}{m_4} = n\frac{h}{m_4}. \tag{2.5}$$

Thus the circulation should be quantised in units of h/m_4. There are some striking types of experiment, discussed below, which verify this simple result, both in concept and in numerical detail. The fact that the quantum of circulation is found to have the value h/m_4 is justification for using the bare ^{4}He mass in equation 2.2.

2.3.3   Quantised vortex lines

Anyone who has let bath water out or who has experienced a tornado has seen a swirling vortex of fluid. So it is reasonable to wonder if such singular rotational events can occur in superfluid helium. And the answer is "yes". The characteristic of a vortex line is that it consists of a core (for example a hole in the bath water) surrounded by circulating fluid. In superfluid ^{4}He, so long as there is a singular core in the fluid, the arguments of the previous section apply. And these arguments tell us that the circulation around the core must be quantised.

What forms the core? In the bath or the atmosphere it is an empty region, one from which the fluid is excluded, essentially by centrifugal forces associated with the rotation. Perhaps that is one way of looking at ^{4}He also, but now we have another way of producing such a singularity in the superfluid. This is evident from equation 2.1. Along our line of singularity we can simply set C, the amplitude of the superfluid wave-function, equal to zero. We then cannot deform a superfluid contour across the singularity, since the phase is indeterminate, meaningless, at this core. The line indeed acts topologically as a hole through the superfluid.

A schematic picture of an isolated vortex line is shown in Figure 2.11. We can deduce a number of things about it. First, consider the velocity field near the line. Using the cylindrical symmetry of a single isolated line, we can see how the velocity varies with radius from the core. In practice, we find that vortices each carry a single quantum of vorticity, i.e. the integer n is +1(or −1). Hence at radius r we have $\oint \mathbf{v}_S.d\ell = v(r).2\pi r$ $= h/m_4$ from equation 2.5. Thus $v(r) = \hbar/(m_4 r)$. The velocity field around a vortex is proportional to $1/r$, in marked contrast to solid body rotation for which $v(r)$ is directly proportional to r.

This result in itself points out the necessity for something different to happen to the wave function near $r=0$. It also gives us a picture of the core's likely size, labelled a_0 in Figure 2.11. At the core radius a_0, the superfluid density C^2 must fall away, to become zero at the centre of the core. One way of looking at this is to say that this is precisely the situation

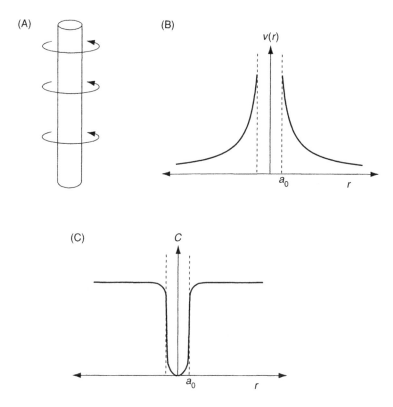

Figure 2.11 A vortex line in superfluid helium. (A) A vortex core with circulation around it; (B) Flow velocity of circulating superfluid near a vortex line (core radius a_0); (C) Wave function amplitude of the superfluid near a vortex line.

we would expect when the velocity rises to exceed the Landau critical velocity v_L. We have already seen that v_L is large, about $58\,\mathrm{m\,s^{-1}}$ at zero pressure. This argument suggests that the core radius a_0 should be approximately $\hbar/m_4 v_L = 0.27\,\mathrm{nm}$. This surprisingly small radius is only of the order of the atomic diameter of a helium atom in the liquid; furthermore a better estimate of a_0 by experiment or theory shows it to be a little smaller even than this, about $0.1\,\mathrm{nm}$. Another way of thinking about a_0 is to equate it with what is called the coherence length ξ of the superfluid wave function. This is the minimum length scale over which the wave function can change significantly, effectively a memory length for the superfluid. These are useful ideas, but must carry a word of caution. This paragraph refers only to low temperatures, where the ⁴He is effectively pure superfluid. As the temperature is raised towards the transition at T_λ the coherence length ξ increases, finally diverging at T_λ where the superfluid disappears.

Finally note that, although a vortex line has helium swirling around it with a quantised circulation, in the bulk of the liquid, equation 2.4 remains true. The bulk of the fluid is still irrotational with curl $\mathbf{v}_S = 0$.

2.3.4 Rotation of superfluid ^{4}He

We had earlier seen that there is a mystery about whether the superfluid component of ^{4}He in a rotating bucket can itself rotate. In practice it does, and the previous section tells us how and why. What happens is that quantised vortices form, with each vortex carrying a single quantum (h/m_4) of circulation. Consider a rotating cylinder of the fluid. The observation is that the meniscus of the rotating liquid is superficially similar to the parabolic shape obtained for a classical liquid like water. However, careful study shows that the liquid surface also contains minute dimples. These are signatures of vortex lines causing local depression of the free surface at the core, just as in the vortex formed in your escaping bathwater.

This picture is supported by various types of experiment. The moment of inertia of the rotating fluid can be measured, verifying that a steady state is reached in which normal and superfluid effectively rotate together with the container. The large scale shape of the meniscus can be observed optically, but the minute vortices need a subtler approach. An imaging technique using negative ions, which are trapped to the vortex core, was used by Richard Packard and co-workers in the late 1970s to observe individual vortex cores as they intersect the free surface. The vortices repel each other slightly and thus form a regular lattice. As expected, the number of vortices increases as the speed of rotation increases. If the rotation (including that caused by the Earth's rotation) is gentle enough, then no vortices are present and indeed the superfluid remains rotationless with respect to the "fixed stars".

The vortex counting method also verifies the magnitude of the quantum of circulation. This had been determined earlier (around 1960) in an experiment by Vinen, in which a rotating cylinder of ^{4}He was fitted with a fine metal wire down its axis. The vibrational modes of the wire are modified by the circulation around the wire in a predictable way. This enabled the value of h/m_4 to be measured experimentally, verifying both the existence of quantised circulation and its magnitude.

2.3.5 Other quantum effects

The existence of a coherent wave function to describe the superfluid has other consequences. One of these concerns the properties of a "weak link" between two areas of bulk superfluid. Such effects were proposed in superconductors by Brian Josephson in the 1960s, and have now been found to have analogues in superfluid ^{4}He.

Josephson's basic idea is that the current I (flow in ^{4}He) through a weak link will be related to the phase difference $\Delta\phi$ between the two reservoirs by the equation $I = I_0 \sin \Delta\phi$, where I_0 is a constant depending on the geometry of the link. Some definitive experiments have been performed by Avenel and Varoquaux in Paris and by Packard and Davis in Berkeley, California. In a helium flow experiment the weak link is a small hole or an array of holes; however "small" here means comparable to the coherence length ξ which we have already seen is of nanometre size. Hence these experiments are not easy. However we now know that the relation between pressure difference (the driving "force") and flow through such a link shows the expected Josephson behaviour. Analogues to the SQUID, discussed in a later chapter, have been constructed and used to detect rotation.

2.3.6 The breakdown of superfluidity

In Section 2.2 we discussed the existence of the Landau critical velocity for the creation of elementary excitations in ^{4}He. Although this velocity of around $60\,\mathrm{m\,s^{-1}}$ can be directly measured by ion experiments, this clean and simple result (e.g. Figure 1.9B) is actually the exception rather than the rule. Consider a simple experiment in which you just push He II through a tube. You find that, although there is indeed superflow at low enough velocities, dissipation starts at a velocity very much lower than v_L. Typically, drag starts to occur at just a few $\mathrm{mm\,s^{-1}}$.

Certainly the explanation for this observation concerns vortices, although any detailed theory still causes controversy and interest. The idea is that the flow generates and sustains a tangled spaghetti of vortex lines, which of course requires an input of energy. One appealing picture due to Schwartz is that of a "vortex mill" where flow causes pre-existing vortex lines to be lengthened continuously by the flow and which are then swept along and tangled down the flow tube. This becomes a plausible idea when you take into account that it is common for a still, well annealed volume of helium to behave quite differently when it is first disturbed by the flow. (It is relevant to note that in the vortex imaging experiments mentioned above, ^{3}He impurity had to be deliberately added to the ^{4}He to aid the destruction of "old" vorticity.) This is a topic of ongoing fascination.

2.4 THERMAL AND MECHANICAL EFFECTS REVISITED

Finally in this chapter, we return briefly to the mathematical description of superfluid ^{4}He. Besides the ideas tabulated above in Table 2.1, there are other important defining equations. These are:

1 *Mass continuity*, giving $\nabla.(\rho\mathbf{v}) = -\partial\rho/\partial t$.
2 *Entropy conservation*, giving $\nabla.(\rho S \mathbf{v_N}) = -\partial(\rho S)/\partial t$.

3 *Thermodynamic gradients acting on the superfluid*, which are simplified by the condition imposed by the phase of the wavefunction being given by curl $\mathbf{v}_s = 0$. This leads to an equation for the acceleration of the superfluid when subjected to pressure and temperature gradients:

$$\frac{\partial \mathbf{v}_s}{\partial t} = S\nabla T - \frac{1}{\rho}\nabla P.$$

4 *Hydrodynamic expressions*, namely Euler's equation, which when linearised for low velocities gives a simple force equation: $\partial(\rho\mathbf{v})/\partial t = -\nabla P$.

These equations give a consistent picture to our understanding of the various strange thermo-mechanical phenomena discussed earlier in this chapter (Section 2.1.3).

For example, the third equation enables us to write down an expression for the "fountain pressure" (Figure 2.5). In equilibrium the superfluid does not accelerate, and we have the simple balance requirement that $\Delta P = \rho S \Delta T$. The entropy per unit mass S is available experimentally from the heat capacity, so the pressure generated is readily calculable, and is found to agree well with the direct fountain effect experiments. It is worth stressing that this is a very large effect in the two-fluid region. At 1.6 K, for example, $S \sim 300\,\mathrm{J\,kg^{-1}\,k^{-1}}$ so that a mere 10 mK temperature difference gives a pressure difference equivalent to a hydrostatic head difference in the ^{4}He of about 30 cm. No wonder the dynamic fountain experiment is so spectacular.

We can also discuss thermal transport in ^{4}He dominated by the normal–superfluid counterflow mechanism. In practice, the flow is often limited by the viscous flow of the normal fluid. Calculation based on the above equations, together with the direct measurements of viscosity from (say) an oscillating disc experiment, gives a satisfactory and consistent description of these apparently different experiments, at any rate for small perturbations from equilibrium. At higher velocities, as in Section 2.3.6, one can observe the onset of further dissipation in a thermal counterflow experiment. Again the onset is at relative velocities of order a few mm s^{-1}. However, there is now classical turbulence in the normal fluid to complicate the picture, as well as the superfluid quantised vortices.

The other major deduction from the equations of the two-fluid model concerns the existence and velocity of second sound. Without reproducing the detail here, it follows that there are two almost independent propagating modes of wave motion. First, there is normal or first sound. This is the usual pressure–density sound wave and has a velocity c_1 given by $c_1^2 = \partial P/\partial \rho$, the usual expression of bulk modulus/density. But because of the loss-free counterflow of normal and superfluid, there emerges also the second sound mode, a propagating temperature–entropy wave. This has velocity c_2 given by

$$c_2^2 = \frac{\rho_S}{\rho_N} \frac{TS^2}{C},$$

where C is the heat capacity. Again it is found that the prediction of the model fits the experimental data very well indeed.

2.5 SUMMARY

To sum up this chapter, let us collect together the salient points about superfluidity in liquid ^{4}He

1   Liquid ^{4}He remains a fluid down to the lowest temperatures, a unique feature of the helium liquids arising because quantum mechanical zero-point energy destabilises the solid.

2   ^{4}He is a boson system, and hence the transition to He II takes place readily (i.e. at a fairly "high" temperature) and bears some similarity to a Bose–Einstein condensation.

3   Superfluidity is possible on the Landau criterion because of the unusual (phonon/roton) dispersion relation.

4   Especially close to T_λ, the transition temperature, a two-fluid model gives a good description of observed phenomena.

5   The superfluid component is characterised by zero viscosity, even though the normal fluid component still exhibits viscous effects.

6   The superfluid is characterised by having a macroscopic wave function with a coherent phase ϕ throughout.

7   The superfluid velocity is related to this phase such that curl $\mathbf{v}_s = 0$. In other words, there is pure potential flow of the superfluid.

8   Because of the coherent phase, circulation is quantised in units of h/m_4, and there can be quantised vortex lines in the superfluid.

9   Many observations are dominated by the behaviour of vortices, so that dissipation often occurs at velocities over 1000 times lower than the Landau velocity.

FURTHER READING AND STUDY

D. R. Tilley and J. Tilley, *Superfluidity and Superconductivity*, Institute of Physics, 1996.

P. V. E. McClintock, D. J. Meredith and J. K. Wigmore, *Matter at Low Temperatures*, Blackie, 1984 (Chapter 5).

J. Wilks and D. S. Betts, *An Introduction to Liquid Helium*, Oxford, 1987.

E. L. Andronikashvili, *Reflections on Liquid Helium*, American Institute of Physics, 1990 (historical reminiscences by one of the pioneers).

The internet references given in Chapter 1 and 5 can also take you to good material on superfluid ^{4}He; or simply try something like "superfluid helium" in a search engine.

PROBLEMS

Q2.1 Summarise the evidence for the two-fluid model in liquid He II. Discuss how such a model is consistent with the requirement that all ^{4}He atoms are identical and indistinguishable.

Q2.2 (a) Show that the angular momentum of a ^{4}He atom circulating around a vortex line is equal to $\hbar$. (b) As stated in Section 2.3.4, "solid body rotation" in a superfluid ^{4}He bath is simulated by an irrotational fluid threaded by quantised vortex lines. In this case, an atom at the edge of a rotating cylinder must rotate with the wall, so that its circulation gives the number of quanta enclosed. Hence estimate how many quanta of circulation would you expect a scientist in Lancaster (latitude 54 °N) to find in a cylindrical bath of helium, diameter 10 cm, as a result of the rotation of the Earth.

Q2.3 In a "modern Andronikashvili" experiment, the resonator is a fine metal wire (diameter 0.10 mm, density 7000 kg m^{-3}) with an attached cylindrical sheath of aerogel (diameter 1.5 mm). Assume that the aerogel is 98% empty space (for the helium, density 145 kg m^{-3}) and 2% silica (density 2200 kg m^{-3}), so finely divided that normal fluid is completely clamped. The resonator vibrates at 240 Hz at 2.5 K. Estimate its frequency at 1 K.

Q2.4 A standing wave resonator tube is set up to observe second sound as in Figure 2.6A. Explain why multiple resonances are detected, and find their frequencies at 1.2 K in a tube of length 5 cm. When the detector is observed synchronously with the generator, explain why, when the generator temperature is at a maximum, the detector indicates cooling for the fundamental and odd harmonics, but heating for the even ones.

Q2.5 (For the more theoretically minded.) Follow up some of the points about thermo-mechanical effects stated in Section 2.4. For instance (a) check the values given for the fountain pressure, (b) work through the derivation of wave equations, including the existence of second sound and the value of its velocity.

Q2.6 Core physics topics to give useful background

(a) Ideal Bose–Einstein gases and the Bose–Einstein condensation
(b) Circulation and the curl operator
(c) The definition and determination of viscosity of a fluid
(d) Damped harmonic motion
(e) Frequency of a torsional oscillator
(f) What determines the velocity of sound?

Q2.7 Possible essay topics.

(a) Thermo-mechanical effects
(b) The dispersion relation in helium

(c) Quantum evaporation experiments using phonons and rotons
(d) Quantised vortex lines
(e) Josephson effects in superfluid helium
(f) The onset of dissipation in He II.

Chapter 3

Experimental techniques

Without attempting to be comprehensive, in this chapter we shall discuss in outline some of the experimental techniques relevant to the study of superfluids. Of particular importance are the topics of refrigeration and of thermometry. The enhanced interest in superfluids following the discovery of superfluidity in ^{3}He (1971) in the mK region has gone hand-in-hand with considerable developments in ultra-low temperature refrigeration. The ^{3}He–^{4}He dilution refrigerator developed at about the same time provides an interesting illustration of the dramatically different properties of the two helium isotopes. We shall stress the interesting physics in these techniques, leaving the (very important!) practical details to the various excellent books on cryogenic techniques, some of which are referenced at the end of this chapter.

3.1 COOLING METHODS

We referred to methods of cooling in Section 1.2 in the discussion of the Third Law. Basically what is required is a "working substance" which shows the following properties:

- The substance has significant entropy at the temperature of interest.
- Thermal equilibrium is reached in a reasonable time.
- The entropy changes not only with temperature but also with some other control parameter.

The first requirement ensures that cooling power is available, so that a passive sample can also be cooled; the second guarantees that it can all happen in a reasonable time scale; the third means that some clever combination of processes can be arranged by which cooling is achieved.

Common working substances include gases (since entropy varies with temperature and pressure), liquid–gas systems (since there is a large entropy difference between the phases, related to the latent heat), ^{3}He–^{4}He mixtures (another two-phase mixture) and paramagnetic solids (since an applied magnetic field is an effective control parameter which changes the entropy).

3.1.1 Cooling to 1 K

This is a well-rehearsed theatre, and the principal actors are helium and nitrogen, as featured in Chapter 1. Specialised machinery and/or commercial suppliers are available for the production of liquid nitrogen and helium. The working substance is the gas itself, and the control parameter is pressure. Basically one starts with gas at high pressure, which is then cooled by allowing it to do work against its surroundings. Liquefaction of helium is not easy, because of its low boiling point, and the final stage of a helium liquefier uses a Joule–Kelvin (Joule–Thomson) expansion, which part liquefies the helium and part generates cold gas which is used in a heat exchanger to precool the incoming helium.

Because of the properties of ^{4}He the temperature range 1–4 K is in many respects the most straightforward to arrange, given a supply of liquid helium. Therefore we consider this range first.

1. ^{4}He evaporation cryostats   The experimental cryostat is simply a glorified thermos vacuum flask (a "Dewar") containing the liquid helium. Even in a vacuum flask, hence with negligible heat leak by conduction, thermal radiation from room temperature can be a significant problem. Remember that the latent heat of helium is very small, so that helium unshielded from thermal radiation derived from a room temperature source causes an unacceptable rate of boil-off. Heat input can be reduced in practice by using liquid nitrogen shielding. Since the power produced by black-body radiation is proportional to T^4, even a single nitrogen cooled wall (and without any silvering to make it reflective) cuts down radiation by a factor of about $(295/77)^4$, greater than 200. Hence most simple research cryostats have a double dewar system with nitrogen in the outer dewar. It is also important to include radiation traps in pumping lines etc. to ensure that there is no line of sight between room temperature and the cold experiment.

Of course it is not always convenient to use liquid nitrogen, especially for helium storage vessels or for superconducting magnet assemblies. Fortunately there is another trick called "superinsulation", an idea to appeal to anyone with a liking for thermodynamics. The idea is simple. Consider an opaque sheet of material sitting in the vacuum between a hot outer surface (temperature T_1) and a colder inner surface (temperature T_2). The sheet will arrive at a thermal equilibrium at temperature T when the heat input to the sheet is balanced by its heat output. The thermal flux from a body of area A and temperature T is given by Stefan's law as

$\varepsilon\sigma AT^4$, where ε is the emissivity ($=1$ for a black body) and σ is Stefan's constant. Hence it is easy to see that the sheet temperature stabilises at a value given by $2T^4 = T_1^4 + T_2^4$. We may note that if T_1 is significantly bigger than T_2, then T is lowered to a fraction $2^{-1/4} = 0.84$ of T_1 and the heat radiated inwards is halved. All this is for a single inserted sheet. To a rough approximation, each sheet inserted cuts down the radiation by a factor of 2. Hence, in practice the vacuum space of a storage dewar is packed with hundreds of aluminised mylar sheets, thin flexible plastic sheeting with metal coating to make it both reflective (reducing ε) and opaque. In this way, the major heat input is down the neck of the vessel, which is often shielded by baffles cooled by the small amount of boil-off helium gas.

Helium baths provide a factor of almost 4 in available temperature, from 4.2 K, the normal boiling point of ^{4}He, down to around 1.2 K by pumping the helium. The lower temperature is a practical limit, limited typically by the size of the pump and by the superfluid film flow up the pumping line. Temperature control is particularly simple, since it can be achieved by controlling the pressure above the helium.

2. Temperatures above 4 K This turns out to be a little more difficult to achieve (surprisingly) than below 4 K, when one needs a stable and controllable temperature. There are several workable methods, some still in use and some dating from the past history of superfluids.

One is to use cryogenic fluids other than helium. However, the next lowest boiling point is that of hydrogen at around 20.4 K, and furthermore it solidifies at about 13 K. By lowering the pressure over the solid temperatures down to about 6 K are attainable, but this still leaves a gap to 4.2 K. None the less, liquid hydrogen was used in small-scale cryostats to cool and then liquefy helium gas. Much of the early work on superconducting lead (transition temperature $T_C = 7.2$ K) and its alloys were carried out using this type of technique. Liquid hydrogen is not of course without its dangers, particularly when combined with liquid air.[1]

The next type of technique is easy but wasteful. It is to have an experimental cell which is weakly coupled to a liquid helium bath at 4.2 K, and to apply controlled heating to the cell. This can be done by having the

---

1 There is an interesting (and not wildly inaccurate) generalisation here, referring to low temperature research in UK around the 1950–60 period. All doctorate students in superconductivity in Cambridge worked on tin ($T_C = 3.7$ K) and indium ($T_C = 3.2$ K) using helium from a central liquefier, with no hydrogen in sight. They claimed that these were ideal superconductors for comparison with theory. On the other hand Oxford students studied lead ($T_C = 7.2$ K) using the "dangerous" liquid hydrogen precooling technique. The author's PhD thesis was on thermal conduction in pure tin and indium, so his educational origins and preconceptions are uncovered!

experiment attached to a copper block in vacuum (an excellent thermal insulator) with a poorly conducting link to the helium bath. A heater on the block is then arranged to respond automatically to a thermometer on the block in order to maintain a controlled temperature.

The best modern solution, popular in recent studies of high temperature superconductors, is to use a "continuous flow cryostat". This makes much more efficient use of the liquid cryogen (helium or nitrogen), as well as providing a more stable temperature. This device operates by drawing the liquid cryogen from a storage dewar in a controlled way using needle valves, often with some additional heating from a temperature controller. The sample can then be in a continuous stream of cold gas at a regulated temperature. Such devices can be arranged to operate at any temperature from about 1.4 K (using a reduced exhaust pressure) to well above room temperature.

Finally, and reverting to the principles of the early days of helium liquefiers which often incorporated experimental access during cool-down, compact closed cycle refrigeration machines are becoming available. These have a small helium gas compressor and allow the gas to cool by doing external work on expansion. Temperature ranges from room temperature down to around 10 K can be achieved without recourse to cryogenic liquids.

3.1.2 Cooling below 1 K

To reach temperatures below 1 K, there are several possible methods. The coldest equilibrium temperatures nowadays are usually reached by a combination of dilution refrigeration for precooling, followed by an adiabatic demagnetisation process using a nuclear spin system. Again, this section will give an outline of the physics involved, leaving much practical detail to the references.

1. Pumped ^{3}He cryostats   Although a cupful of liquid ^{4}He costs only about the price of a cup of beer, a cupful of liquid ^{3}He costs about the same as a new motor car. Therefore, it is only used sparingly and in a closed system, precooled by liquid ^{4}He. Nevertheless a simple ^{3}He refrigerator or cryostat is a useful tool, since temperatures of around 0.25 to 0.3 K can be attained with comparatively simple equipment and good cooling power. There is no problem of superfluid film flow in a vessel of liquid ^{3}He, so that the vapour pressure attainable by pumping is much less than that reached in a similar vessel of liquid ^{4}He.

2. Pomeranchuk cooling   This curious cooling method, named after its inventor, is interesting for two reasons. One is historical, since it was during its operation in 1971 by Lee, Osheroff and Richardson that superfluidity was discovered in liquid ^{3}He, as already mentioned in Section 1.3.2. The other, also mentioned in Chapter 1, is that it uses a cute bit of thermodynamics, based on the negative slope of the liquid–solid line on the phase diagram of pure ^{3}He (see Figure 1.4). The fact that the solid (through its random spins) is more disordered than the liquid (in which the spins are ordered by Fermi statistics) means that conversion of liquid to solid by applying pressure produces cooling. This method only works below 0.3 K, and it has a low temperature limit around 1–2 mK, below which temperature the spins in the solid become ordered. Furthermore a clever indirect mechanism (usually using ^{4}He and a bellows arrangement) must be devised to apply the appropriate pressure, since application of pressure directly down a fill line from room temperature is thwarted by blockages in the tube above 0.3 K. This cooling method is now superseded as a practical proposition by the techniques which follow.

3. ^{3}He–^{4}He dilution refrigeration   The dilution refrigerator is now the machine of choice for experiments below 1 K. It is well described elsewhere[2], but it is worth consideration here as a beautiful combination of engineering and basic physics. The physics involved is a fascinating illustration of the difference in quantum properties between odd (fermion) and even (boson) particles. We have already noted in Chapter 1 that the helium liquids ^{3}He and ^{4}He are entirely different in their properties at low temperatures. So what happens if we cool a mixture of the two isotopes?

The answer is that there is a phase separation below about 0.8 K. This is the only known example of isotopic ordering. The coexistence phase diagram is illustrated in Figure 3.1.

Consider the fate of a vessel containing a roughly 50/50 mixture of the two isotopes cooled from a high temperature. Above about 1.3 K, the liquid exists as a normal fluid of ^{3}He and ^{4}He in solution. Then at point A in the figure it goes through the λ-line and becomes a ^{4}He-like superfluid. It has properties similar to those described in Chapter 2, but with an enhanced normal fluid density arising from the ^{3}He in addition to phonons and rotons. It is still a single-phase uniform solution. Then at point B, about 0.8 K, a phase separation occurs, with a new ^{3}He-rich phase starting to form at point B'. As the temperature is decreased further, the difference between the two phases gradually widens until at very low temperatures we have the two phases labelled C and C'. In the trade, these are called the dilute phase and the concentrated phase respectively,

2 See Lounasmaa's book for a full discussion. An example of a modern implementation in the author's laboratory is found in Journal of Low Temperature Physics 114, 547–570 (1999).

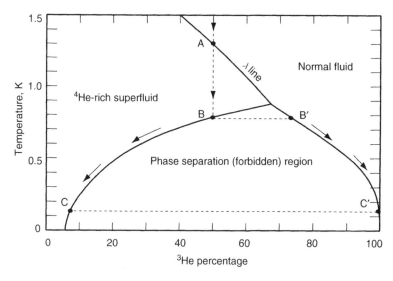

Figure 3.1 Properties of ^{3}He–^{4}He isotopic mixtures.

referring to the ^{3}He concentration. In a gravitational field, the lighter concentrated phase sits on top of the heavier dilute phase.

Now an outstanding feature of the two phases is that, although at the lowest temperatures the concentrated phase at C′ is virtually pure ^{3}He, the dilute phase still contains about 6% ^{3}He. This is fortunate for dilution refrigeration, which relies on ^{3}He being soluble in ^{4}He as we shall see below. The reason is not difficult to fathom. A helium atom, ^{3}He or ^{4}He, is attracted by van der Waals forces to all other helium atoms. We have already noted (Section 1.3.1) that helium is liquid because of quantum zero-point energy. Now because ^{3}He is much lighter than ^{4}He, its zero-point motion is even more violent, leading to the liquid being considerably less dense than liquid ^{4}He (in fact there is a 28% difference in atomic volume between the two). Hence a single helium atom will be more strongly attracted to be in ^{4}He than in ^{3}He. This explains without more ado why the concentrated phase is pure, since any single ^{4}He atom will prefer to be surrounded by the more dense helium of the dilute phase. This simple argument works because in the Bose ^{4}He superfluid, no other energies are involved. Similarly a single ^{3}He atom also likes to be in a ^{4}He environment, and this is why the solubility is not zero. The solubility is not complete, however, because ^{3}He is a Fermi–Dirac system, requiring additional ^{3}He atoms to be added with a non-zero kinetic energy equal to the Fermi energy. The 6% solubility limit thus arises from the balance between the negative binding energy of a ^{3}He atom and the positive Fermi energy which depends on the ^{3}He density in the dilute phase.

Back to the physics of dilution refrigeration. At temperatures much lower than the Fermi energy E_F, the entropy of a simple Fermi gas is given by

$$S = \frac{\pi^2}{2} N k_B \frac{k_B T}{E_F}. \tag{3.1}$$

It is linear in T and is smaller than a Maxwell–Boltzmann gas by a familiar factor of order $(k_B T/E_F)$. Now the concentrated phase has a much higher Fermi energy than the dilute phase (roughly by a factor of $(100/6)^{2/3} = 6.5$). Hence if we can persuade ^{3}He atoms to go from the concentrated to the dilute phase, then they go from a low entropy (ordered) phase to a higher entropy (more disordered) phase. To do so, they must absorb heat (equal to $T\Delta S$) and hence cool their surroundings. This is analogous to the more familiar gas–liquid cooling where molecules leave the ordered liquid phase to the more disordered gas, thereby absorbing latent heat $T\Delta S$ from the surroundings. In our case, the cooling power of the process is thus proportional to T^2 (since ΔS is itself proportional to T) and to the rate $\dot{n}_3$ mol s^{-1} at which ^{3}He atoms pass through the phase boundary. In practice, the cooling power is given by

$$\dot{Q} = 84\dot{n}_3 \left(T^2 - T_0^2\right) \tag{3.2}$$

where T_0 is the base temperature of the refrigerator, limited by the circulation rate and by conduction as mentioned below.

The practical implementation of a ^{3}He–^{4}He dilution refrigerator is sketched in Figure 3.2. The volumes of ^{3}He and ^{4}He are arranged so that the phase boundary occurs in the mixing chamber. Movement of ^{3}He through the boundary is generated by applying heat at the still and pumping on it. This age-old method of distilling mixtures ensures that the more volatile ^{3}He is evaporated rather than ^{4}He, which has a much lower vapour pressure at still temperature, typically 0.6 K. Hence ^{3}He is removed from the dilute phase at the still and is then replenished by solution ("dilution") at the mixing chamber. We apply heat at the still and get cooling at the mixing chamber. The machine is turned into a continuously operating refrigerator by returning the ^{3}He gas from the pumping system, condensing it back to liquid at a 1 K ^{4}He pot and returning it via heat exchangers to the mixing chamber. The success of this whole operation depends on the efficiency of these heat exchangers and the design of the tube sizes, bearing in mind that the moving ^{3}He is a highly viscous substance in either phase at these temperatures. The ultimate useful temperature T_0 reached depends on a balance between various factors, but

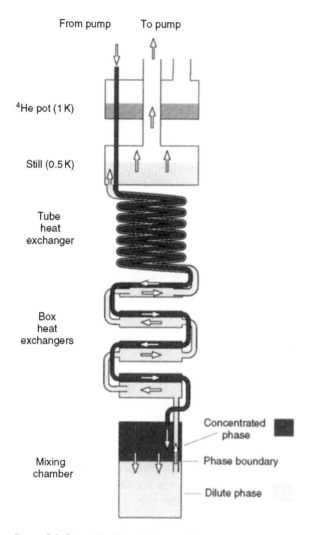

From pump   To pump

⁴He pot (1 K)

Still (0.5 K)

Tube
heat
exchanger

Box
heat
exchangers

Mixing
chamber

Concentrated
phase

Phase boundary

Dilute phase

Figure 3.2 Principle of the dilution refrigerator.

basically the cooling power goes down as T^2, whereas the viscosity of the Fermi fluid rises as T^{-2} (see Section 5.3.5). Our recent refrigerator at Lancaster University, referenced in Figure 3.2, can reach around 2 mK in continuous operation over several months.

4. Adiabatic demagnetisation   Since the superfluid transition temperature of liquid ³He is in the range 1–2 mK, a further stage of cooling is needed.

Nowadays the preferred technique is to employ an adiabatic demagnetisation stage precooled by a dilution refrigerator.

The technique of adiabatic demagnetisation has a long history of success. The method was developed in the 1930s, well before enough ^{3}He gas became available for that to be considered as a practical proposition. The idea is well covered and discussed in almost any text on thermal or statistical physics, in addition those specifically on the low temperature techniques. The important thing to appreciate here is that the technique does in fact work as simply as the textbooks imply! It is a beautiful example of thermal physics in action.

The idea can be illustrated by considering an ideal spin 1/2 solid of N atoms. By "ideal" one means that the solid contains N localised spins (shorthand for magnetic moments in this context) which are weakly interacting. Hence the solid is paramagnetic. The spin 1/2 solid has just two states for each magnetically active atom (spin up and spin down) which makes it easy to visualise, but there is no problem about generalising the discussion to any quantum number. In an effective magnetic field B, the spin states have energies of $+\mu B$ and $-\mu B$, where $-\mu$ and $+\mu$ are the appropriate z-components of the magnetic moment. Hence in thermal equilibrium at temperature T, we can use the Boltzmann distribution immediately to find the occupation numbers of the two states and hence the thermal properties. As is well known, the entropy curves are as illustrated in Figure 3.3.

At high temperatures (meaning $k_B T \gg \mu B$), the spins are completely disordered, giving an entropy of $N k_B \ln 2$ since there are the two possible spin states. As the temperature is lowered below $\mu B / k_B$, the spins favour the lower state, lowering the entropy towards zero. The entropy is a function of $\mu B / k_B T$ only. Figure 3.3 shows the entropy at two different magnetic fields, and illustrates the principle of the cooling method. First a large magnetic field is applied, and the solid is cooled by connection to a precooling refrigerator for as long as patience allows. The spins align in the field and the entropy is reduced by this precooling process, illustrated by passing from points P to Q on the graph. Next comes the "adiabatic demagnetisation". The system is thermally isolated (adiabatic), and the field is reduced (demagnetisation). Adiabatic implies that there is no way for the spin configuration to change, rather the degree of alignment remains the same. Hence the ratio $\mu B / k_B T$ remains constant. This means that, as B reduces, so does the thermal energy scale and T is lowered. The process goes from points Q to R in the figure. Finally we may note that the cooling capability of the spin system may be seen from Figure 3.3, since the shaded area is $\int T dS$ – the heat input required to go from point R to point R'.

Before leaving this topic it is worth making a few practical remarks.

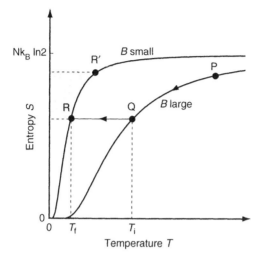

Figure 3.3 Adiabatic demagnetisation of a spin 1/2 solid. P to Q shows precooling in a high field. Q to R shows cooling by adiabatic demagnetisation. R to R′ indicates the cooling power available.

Choice of coolant – precooling The first requirement is that we satisfy $\mu B_i \sim k_B T_i$ where B_i is the initial (high) magnetic field available and T_i is the temperature attainable with the precooling stage, indicated by point Q on Figure 3.3 refers. Before the advent of the dilution refrigerator, precooling was usually limited to a 1 K ^{4}He bath. The condition is then satisfied for a typical electronic magnetic moment with a 1 T field, readily attainable with an electromagnet at room temperature. Temperatures as low as 2–10 mK can be attained in this way. However, the dilution refrigerator revolution changed all that, since the same temperature range became attainable in continuous operation. Nowadays, the microkelvin range is reached using nuclear spins (roughly a thousand times smaller), precooled to about 5–10 mK (a hundred times larger) in the field of a superconducting solenoid (almost 10 times larger).

Choice of coolant – final temperature The final temperature reached is simply $T_i(B_f/B_i)$, where B_f is the final (small) field. This is point R on the graph. However this simple statement hides a few subtleties. For a start, one is tempted to set $B_f = 0$ and hence to attain the absolute zero, in contradiction to the third law of thermodynamics. But of course B_f is in fact not simply the applied field, but μB_f characterises the final splitting of the energy levels of our spin system. This is never zero, because of interactions between neighbouring spins, even when the applied field is zero. The effective B can be written to a good approximation as $\sqrt{(B_a^2 + B_{int}^2)}$, where B_a is the applied field and B_{int} is the effective interaction field sensed by a spin. Hence the minimum attainable temperature becomes $T_i(B_{int}/B_i)$. It is this internal field which limits the utility of atomic (electronic) spin systems. The substances used are typically highly diluted salts, with only a few active atoms surrounded by a lot of non-magnetic padding. The lowest temperatures are obtained with "CMN", Cerium Magnesium Nitrate with plenty of water of crystallisation, in which the only magnetic ions are provided by the cerium. But the dipole-dipole interactions still produce ordering, and hence a limit to the final temperature, at about 1 mK. Actually there is a Murphy's Law problem here. As Figure 3.3 illustrates, the lower the final effective field, the lower also is the available cooling power. In old experiments, the best choice of paramagnetic salt was often dictated by a compromise between low temperature and high cooling power. The same constraints are relevant in cooling by nuclear spins. In this case, the refrigerant is usually demagnetised to a substantial final field, so that the cooling power is kept at a useful value, albeit at much higher nuclear spin temperature than the minimum attainable.

Conditions for cooling with nuclei   The common choice of refrigerant here is copper metal. Both ^{63}Cu and ^{65}Cu are spin 3/2 nuclei in a cubic environment which ensures that the interaction effective field is very small. (Remember that the interaction energy of two spins a distance r apart is of order μ^2/r^3, and nuclear spins are about 2000 times smaller than electronic spins.) Hence temperatures into the n K range are in principle possible. However, it is here that questions of thermal contact and of heat leaks become important. This is illustrated in Figure 3.4, a schematic diagram showing the problems encountered in cooling a sample such as liquid ^{3}He. There are two bits of seriously bad news to frustrate the exercise. These are indicated by the thermal resistances R_1 and R_2. The resistance R_2 arises from poor thermal contact between the copper coolant and the ^{3}He

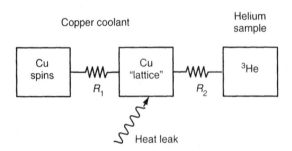

Figure 3.4 Schematic diagram showing the problems for cooling a helium sample using nuclear cooling of copper.

sample. This is often called the Kapitza resistance, a boundary resistance arising from a mismatch in acoustic properties between helium and a solid. In practice this resistance is made as small as possible by having a large surface area of contact, arranged by the use of very fine sintered sinter powder (typically with grains as small as 70 nm), the same technique as used in the heat exchangers in a dilution refrigerator. As if that is not bad enough, there is another effective resistance R_1 between the "lattice" (not distinguished from the conduction electrons in this jargon) of the copper and the nuclear spins. This is characterised by a spin-lattice relaxation time τ_1, which is proportional to $1/T$ according to the "Korringa relation". (The $1/T$ dependence comes from Fermi statistics again, because of the increasing difficulty of finding full initial and empty final states for a spin-lattice relaxation which must dump energy of $2\mu B$.) Copper is chosen because of its reasonably fast relaxation time, but even that is of the order of 1000 s at 1 mK. The final piece in this story is to realise that heat leaks, impossible to eliminate entirely, are certain to enter the helium or the copper lattice directly, so that they are bound to leave the sample at a higher temperature than the copper nuclei, since they have to flow to the refrigerant through one or both of these thermal resistances.[3] The final sample temperature is thus determined by heat leaks and thermal contact and not simply by the nuclear spin temperature. Using this logic at Lancaster University, we have developed methods for cooling liquid ³He deep into its superfluid phases (to about 80–100 μK) for times lasting several days. The arrangement is sketched in Figure 3.5.

Heat switches One final practical point arises from Figure 3.5. An adiabatic demagnetisation cooling is a form of single-shot cooling which requires a heat switch. In the precooling phase, good connection to a refrigerator is needed, whereas in the adiabatic leg of the operation (and after), the cold stage must be isolated from the precooling refrigerator. A heat switch is needed. For the paramagnetic salts of old, helium exchange gas was often used, but of course this leaves an undesirable superfluid film. Mechanical switches are cumbersome and can cause heating when activated. The modern elegant solution is to use in the thermal link a piece of pure (type I – see below) superconductor, such as aluminium. Aluminium has a superconducting transition temperature T_C of about 1 K.

---

3 Perhaps a simple analogy helps here? If you wanted to evacuate your room, what would you do? By "evacuate", I mean remove the air (not the occupants – I hope they left first !) from the room to as low a pressure as possible. Getting the pump with the lowest possible ultimate vacuum is not remotely important. But sealing the cracks is vital – like keeping down the heat leak in our situation. And the other trick is to get a wide connecting pipe (lowering the resistance to the coolant) to a large throughput pump (having a refrigerant with a large cooling power).

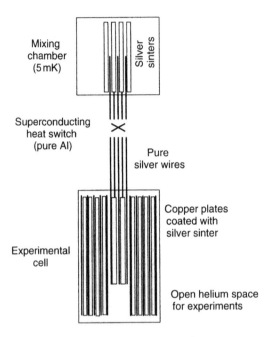

Figure 3.5 A method for cooling liquid ^{3}He.

In its normal (i.e. non-superconducting) state, the pure aluminium is an excellent thermal conductor; and even at millikelvin temperatures it remains normal in a modest (100 mT) magnetic field. However, when it becomes superconducting, well below T_C, there are virtually no normal electrons to carry heat and the phonons are also very few, so that it becomes effectively a thermal insulator. Hence a magnetically-operated heat switch can be readily incorporated, a very neat application of the basic properties of superconductivity. (Yes, that is correct. A superconductor is an ideal conductor of electricity, but an almost ideal insulator for thermal conduction. Read more in the next chapter!)

3.2 THERMOMETRY AND THERMAL CONTACT

This section can be fairly short. The problems of thermal contact have already been touched upon in the previous section. Thermal contact generally becomes a more acute problem as the temperature is lowered. It is important to remember that the temperature of a thermometer tells you the temperature only of that thermometer, and not necessarily that of your sample. Because of the Kapitza boundary resistance, for instance, a thermometer immersed in a helium bath will only faithfully register the bath temperature if it is above about 100 mK, although this depends on the power dissipation (deliberate or accidental) in the device. Similarly a sample immersed in the bath may not attain the bath temperature.

There is a nice little theorem about cooling and thermometry. This is that any technique which forms the basis of a cooling method can also be used as a thermometer in the same range. The idea is pretty. Cooling, as we have seen, requires a working substance whose entropy varies with temperature and with another control parameter. Hence its entropy and other equilibrium properties are necessarily changing with temperature, so that it has potential as a thermometer.

It is often useful to distinguish between primary and secondary thermometers. Primary thermometers are used to connect to the absolute Kelvin scale directly; secondary thermometers are convenient devices calibrated against such primary ones. An internationally agreed and transferable temperature scale is maintained by institutions such as the National Physical Laboratory in UK; "ITS-90" gives fixed points between 1 K and 1600 °C and is in constant review as to its consistency with thermodynamics and its possible extension to lower temperatures.

3.2.1 Primary thermometers

A primary thermometer must have a simple theory to connect it to basic thermal physics and hence to the Kelvin scale. There are three useful candidates here.

1. Gas thermometry   A simple result of statistical physics is that an ideal Maxwell–Boltzmann gas obeys the equation of state $PV = nRT$. And helium gas at low pressure is almost an ideal gas. Hence gas thermometry can be used as a primary thermometer to connect low temperatures to the Kelvin scale, with its one fixed point at the triple point of water. Of course this cannot be extended to extremely low temperatures, since interactions even in helium become important. After all helium liquefies. However, at low enough pressure it is just about feasible down to about 3 K.

2. Paramagnetic systems   Again, there is a simple theory for an ideal paramagnetic system. Boltzmann statistics apply to the localised spins, and the energy splitting is governed by the applied magnetic field. Our discussion of adiabatic demagnetisation was based on these ideas, and in the region where spin-spin interaction energies are negligible, this is an excellent primary system. At high temperatures ($k_B T \gg \mu B$), Boltzmann statistics show that the magnetic susceptibility χ of the substance obeys Curie's Law, namely $\chi \propto 1/T$. Using CMN, the scale can be joined on to the gas scale and extended down towards the mK range.

There are difficulties in practice, in that susceptibility contributions can arise from other magnetic substances in the neighbourhood. A very clever way around this is to study nuclear spins using NMR, a method which works from tens of mK down. The favourite substance here is ^{195}Pt, a simple spin 1/2 nucleus. The strength of the NMR signal is proportional to the nuclear susceptibility, i.e. to the Curie's law. The method has the advantage of simply measuring the relative population of the two specific resonating spin states, so it is unaffected by other magnetically active spins. There is the further advantage that it is in principle possible to measure the spin-lattice relaxation time τ_1 by NMR, an independent check and a calibration against the Korringa constant ($=\tau_1 T$) for the nucleus assuming that this is known. It is relevant here to repeat the sermon that this technique tells you the spin temperature of the nuclei, which may or may not be the same as the temperature of your sample (see again Figure 3.4, replacing Cu by Pt).

3. Noise thermometry   Another potentially primary thermometer is noise thermometry. Unlike the previous two methods, this is not related directly to a cooling method. It is however based on fundamental thermal physics and the principle of equipartition of energy, thanks to ideas originated by Einstein relating fluctuations to dissipation. You simply measure the voltage noise (Johnson noise, effectively the Brownian motion of electric charges) generated in a resistor of resistance R at temperature T. The result is that the RMS noise measured in a frequency bandwidth Δf is given by

$$\langle V^2 \rangle = 4k_B T R \Delta f. \tag{3.3}$$

Nowadays there are very sensitive methods of measuring this RMS noise using superconducting SQUID circuits, and this method can be used to rival, check and even improve upon the NMR method, particularly since it measures the lattice temperature rather than the nuclear spin temperature (compare Figure 3.4).

3.2.2   Secondary thermometers

There are many possible secondary thermometers, useful in various temperature ranges. We mention a few.

1. Phase diagrams of helium   These provide very convenient secondary thermometry in a number of ways. The vapour pressure curves for ^{4}He and for ^{3}He are well known and tabulated, and these provide by far the most convenient reproducible scale in the 0.5–4.2 K range. Simply pumping down a bath of ^{4}He and measuring the pressure above the liquid gives a remarkably useful measure of the temperature. A little caution is needed on warming up a helium bath, since when once T_λ is exceeded, the bath is unlikely to be at a uniform temperature.

The other useful phase diagram is the melting curve of ^{3}He. The pressure in a cell containing a solid–liquid mixture is used to track the temperature. Note that this method relates to Pomeranchuk cooling.

In addition, these phase diagrams can provide useful and reproducible fixed points in relation to the superfluid phases. ^{4}He has the λ-line, a very clear indication. The phase diagram of superfluid ^{3}He is universally used as a reference point for sub-millikelvin work, even though the actual Kelvin numbers have changed quite a lot since 1971. Physicists working in this range nowadays have just about agreed on a working scale, i.e. what numbers in kelvins should be attached to the transition lines.

2. Resistance thermometry   This is a subject in itself. At temperature above about 20 K it is possible to use the resistivity of a pure metal as a guide, with the platinum resistance thermometer a continuing favourite among the standards laboratories of the world. Overlapping with that and extending to the mK range, one can use carbon or germanium thermometers. Some sorts of obsolete carbon composition resistors are still useful, although they are not so easy to find with the advent of temperature-insensitive metal-oxide or metal film resistors. (After all in an electronic circuit you would rather have your resistor to be stable!) The resistance of the semiconductor rises rapidly as temperature is lowered, so it is possible to find very sensitive devices in a limited temperature range. A major disadvantage for this method is that it is necessarily dissipative. It becomes difficult to make effective thermal contact to a carbon or germanium device below about 10 mK.

Other methods

Thermocouples can be useful, particularly because certain alloys with magnetic impurity (e.g. Au + 0.03% Fe) exhibit giant thermoelectric effects at low temperatures. Various specific alloys (e.g. Pd + Fe) have resistive anomalies also which can be used into the mK range, but all these effects are sample-dependent. Other electrical properties such as the characteristic of a diode are useful above 4.2 K.

Superconductors provide useful fixed points down towards 10 mK, since their values of T_C are well-known and reproducible, so long as magnetic fields are excluded.

3.3 SUMMARY

In this chapter we have looked at the physics behind various low temperature techniques.

1 Cooling with helium, either in a closed cycle refrigerator or as a liquid enables temperatures to be reached down to 1 K.

2 Below 1 K the ^{3}He-^{4}He dilution refrigerator enables one to attain stable temperatures to almost 2 mK.

3 To cool below this, the favoured method is to use the adiabatic demagnetisation of copper nuclear spins.

4 Compromises must be reached between ultimate temperature and cooling power in this technique. Reducing heat leaks is paramount because of poor thermal contact.

5 Primary thermometers relate to reliable basic thermal physics. Perfect gases and paramagnets are very useful, and so is noise thermometry.

6 Secondary thermometers give transfer and portability for temperature measurement.

7 At mK temperatures the ^{3}He superfluid phase diagram provides a useful transfer between laboratories.

FURTHER READING AND STUDY

Guy White and Philip Meeson, *Experimental Techniques in Low Temperature Physics*, 4th edition, Oxford, 2002.

O. V. Lounasmaa, *Experimental Principles and Methods below* 1 K, Academic Press, 1974 (A classic work, still very useful).

P. V. E. McClintock, D. J. Meredith and J. K. Wigmore, *Matter at Low Temperatures*, Blackie, 1984 (Chapter 7).

G. R. Pickett, *Microkelvin Physics* in *Reports of Progress in Physics*, Vol. 51 pp. 1295 ff, 1988.

PROBLEMS

Q3.1 Explain, with the aid of temperature-entropy diagrams why the Third Law implies that it is impossible to reach the absolute zero. [Revise Chapter 1.2.2 if necessary.]

Q3.2 The cooling power of a dilution refrigerator at temperature T is often written as $\dot{Q} = K\dot{n}_3 T^2$, where $\dot{n}_3$ is the ^{3}He circulation rate. Derive this expression, and from the simplifying assumptions made in your derivation, discuss the circumstances in which it is useful. What determines the constant K?

Q3.3 Expand on the idea introduced in Section 3.2 that a cooling method can also form the basis of a method of thermometry. Give examples. Is the reverse hypothesis true?

Q3.4 In some set-ups, it is required to run a Pt NMR thermometer in the magnetic field which forms the final field of an adiabatic demagnetisation stage. If the NMR coil resonates at 50 kHz, a final temperature of 5 µK is required and the precooling field is 7 T, what is the highest acceptable precooling temperature? [^{195}Pt has an NMR frequency of 9.2 MHz T^{-1}.]

Q3.5 Describe the features required in a substance suitable for cooling by adiabatic demagnetisation, suggesting why copper is the common choice. Why are paramagnetic salts not used for cooling ^{3}He into the superfluid state?

Q3.6 Core physics topics to give useful background

   (a) Temperature scales
   (b) Black body radiation
   (c) Entropy of an ideal Fermi–Dirac gas
   (d) Statistical physics of an ideal spin 1/2 solid

Q3.7 Possible essay topics.

   (a) Reaching the lowest temperatures
   (b) Pomeranchuk cooling
   (c) Dilution refrigeration
   (d) Adiabatic demagnetisation
   (e) Thermometry below 4 K
   (f) Problems of thermal contact

Chapter 4

Superconductivity

This chapter is aimed at describing the physics of superconductivity without any attempt at being comprehensive. This is because the whole subject covers vast areas of experimental physics, materials science, theoretical physics, applied device physics and engineering. We shall concentrate principally on features of superconductivity which resonate with other parts of the physics of superfluids.

4.1 THREE BASIC PROPERTIES OF SUPERCONDUCTORS

In this section, we take a rapid tour through the history of the subject. Many of the important ideas about what makes a superconductor are included under the following three headings.

4.1.1 Zero electrical resistance

As already mentioned in Chapter 1, the discovery of the first superfluid state was by Kamerlingh Onnes in 1911, soon after his liquefaction of helium. Since the electrical resistivity of a pure metal was known to reduce as its temperature is lowered, he was making measurements of the resistance of a sample of mercury, which could be purified by distillation. What he found was totally unexpected. The resistance dropped abruptly to zero as indicated in Figure 4.1, at a transition temperature T_C which happened to be close to 4.2 K. It was for this reason that the term "super-conductor" was coined – the mercury was a superconductor.

Any statement about "zero" needs to be made carefully. Of course the original result was simply that the resistance was very much less than that in the normal state just above T_C. But later experiments have set almost astronomical limits on the resistance ratio. Truly zero resistance means that a current induced in a loop of the material would never decay. A simple laboratory experiment to investigate this is to cool a loop of the

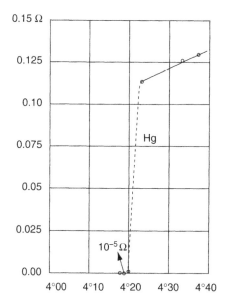

Figure 4.1 The resistivity of mercury around 4.2 K. The discovery of superconductivity by Kamerlingh Onnes.

Sources: From Kittel, C., Introduction to Solid State Physics, 7 E. Copyright 1995 John Wiley & Sons, Inc. Reprinted by permission of John Wiley & Sons, Inc.

wire in a magnetic field through the transition temperature, and then to remove the source of the magnetic field. This removal then causes a current to flow around the loop, maintaining the same magnetic flux through the loop. For a resistive loop having resistance R and inductance L, this current (and hence the magnetic fields produced by it) will decay with a time constant equal to L/R. In practice nobody has observed decays in this sort of experiment, even after a wait of several years, and an upper limit on the DC resistivity of a superconductor is about 10^{-24} of its normal state value.

One might note that this type of "experiment" is performed regularly by users of superconducting magnets, such as the magnets in large NMR machines in hospitals or laboratories. Such magnets need to be stable over a period of years on a level typically better than 1 in 10^6. This stability is routinely achieved by placing a thermally operated persistent switch across the magnet leads as illustrated in Figure 4.2. The switch, between point A and B, consists of a length of superconducting wire in intimate contact with a heater. The magnet is usually immersed in liquid helium to keep it cold. When the magnet is first installed and cooled down, the NMR system is tuned up by the engineer by adjusting the current from

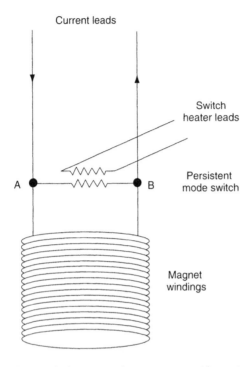

Figure 4.2 A superconducting magnet with persistent mode switch.

a power supply, with the persistent switch heater on so that the current passes from A to B through the superconducting (zero resistance) magnet rather than through the switch. When the field satisfies the NMR require-ment, the switch heater is turned off, and after a few seconds the current from the power supply is ramped down to zero. The final result is then that the field in the magnet remains the same for as long as you care to wait; the current continues from A to B in the magnet and returns from B to A through the switch element. And the engineer can go home, taking his power supply with him! One might add that if he is a sensible fellow, he will also thoroughly disconnect and hide the leads to the switch heater, since any excitation of them would add resistance to the otherwise superconducting circuit; and this would cause not only the loss of the required NMR field, but also an exciting catastrophe since all the energy stored in this field would be dumped in the circuit, doing no good at all to the liquid helium bath surrounding the assembly.

Let us recognise here that not all materials are superconductors. A rough guide is that elemental metals (like Kamerlingh Onnes's mercury) have low transition temperatures in the kelvin range (Nb is the highest at

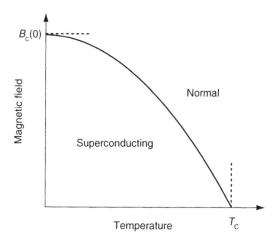

Figure 4.3 The phase diagram of a typical superconductor.

around 9 K), and that by and large superconductivity appears favoured by metals with the lowest, not the highest conductivity. Monovalent metals are never superconductors, at any rate under normal conditions. The same paradox is valid for modern (high temperature) superconductors also. The exotic substances which have T_C values at 90 K or even higher are layered materials teetering on the brink of being insulators rather than metals at all. A typical superconducting magnet as mentioned above has wire fabricated from Nb alloys, notably a well-studied NbTi alloy with a T_C value over 12 K and a high current-carrying capacity.

When superconductivity was discovered, it opened up all sorts of practical possibilities, since electric currents without power losses are an engineering dream. It took 50 years before any of these dreams (like the magnet just mentioned) became at all realistic. This is because it was found quite early on that a magnetic field destroys the superconducting state. You can draw a B–T phase diagram for each substance, as illustrated in Figure 4.3. Superconductivity only occurs at low temperatures and at low magnetic fields. Roughly speaking, many superconductors follow a "law of corresponding states" in that the critical field B_C scales linearly with the transition temperature T_C, i.e. the phase diagram has a single scaling parameter specific to the substance. However, it is important to realise that this "law" is only a rough guide.

Hence, even in zero applied magnetic field, the current carrying capacity of a superconducting wire is limited. This is because the current itself generates a magnetic field and that field will cause the wire to become normal and hence resistive. We should also note that some resistance is observed at high frequencies, so that the property summed up by the simple statement

$$R = 0 \qquad (4.1)$$

is only strictly valid for DC measurements and for low applied magnetic fields.

It turns out that the influence of a magnetic field on a superconductor is of basic importance to our understanding. I guess that until about 1950 superconductivity was studied in only a handful of laboratories, but it is surprising that the strangeness of the magnetic properties of a superconductor was not discovered until the work of Meissner and others in the 1930s, 20 years or so after the discovery of zero resistance.

4.1.2 The Meissner effect

What Meissner showed in 1933 is that there is much more to superconductivity than perfect conductivity.

Consider a lump of superconducting metal cooled below T_C in zero applied magnetic field. If we then apply a magnetic field to it, then Maxwell's equations (or commonsense?) tell you that the B-field inside the superconductor has to remain zero. This is because, if the resistivity is zero then there can be no electric field E in the bulk of the material since that would imply an infinite current. Hence since $E=0$ everywhere Maxwell tells you that

$$\frac{dB}{dt} = 0. \qquad (4.2)$$

What happens is that persistent surface eddy currents are set up in the lump of superconductor to maintain the status quo. This conclusion is indeed in accord with experiment.

However, consider now the sort of experiment illustrated in Figure 4.4, when the same experimental constraints are reached in a different order. Let us now first apply the B-field above T_C and then cool the lump of material below T_C. The result expected from Maxwell's equations alone is again easy to derive and is indicated in Figure 4.4A. Above T_C, the (non-magnetic) metal will be penetrated uniformly by the B-field. When it is subsequently cooled below T_C, for reasons just stated, the perfect conductor maintains exactly the same field pattern, to ensure again that $dB/dt = 0$. Easy! But proved wrong by experiment!

What in fact happens is much more remarkable. The superconductor actively expels all magnetic flux from the bulk, as illustrated in Figure 4.4B.

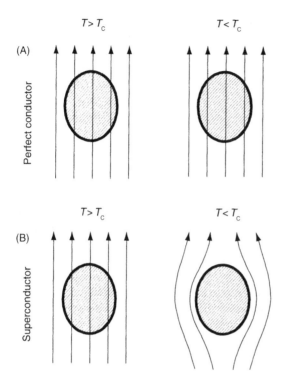

Figure 4.4 The Meissner effect, showing magnetic field lines near a superconductor. Flux is actively excluded as the material becomes superconducting. (A) "Perfect" conductor; (B) Superconductor.

This is called the Meissner Effect, and can be summed up by the new equation

$$B = 0. \tag{4.3}$$

This magnetic flux exclusion is again performed by surface eddy currents being set up, but now they are set up actively without any change in external applied field.

The Meissner effect is now understood to be a basic property of the superconducting state. It is particularly interesting to note that it implies a reversible transition, in the sense that the order of application of cooling and field makes no difference; the final state of the superconductor is the same (as in Figure 4.4B below T_C). The same point is made if we roll on the camera in Figure 4.4 and work out the state if, after cooling, we remove the applied B-field again. The superconductor simply winds down its surface eddy currents, to leave the material in the same condition as it would have been if cooled through T_C in a zero field. On the other hand, the perfect conductor would only now set up eddy currents to resist the field change, and it would be left as a large magnetic dipole with eddy currents circulating for ever.

A well-known demonstration of the Meissner effect is to indulge in a little magnetic levitation. A permanent magnet near a superconductor will induce surface currents in the superconductor to exclude the applied B-field. This magnetic response is thus a very strong one and one which opposes the applied field. Hence a magnet can be made to float over a dish made of a superconductor, a "Maglev" effect that has been used in Japan to produce levitation and hence the low-friction suspension of a train above its track.

The existence of the Meissner effect immediately lets us understand why a magnetic field destroys the superconducting state. It is all a matter of minimising the free energy of the system. In zero applied field, there is clearly an advantage in forming the superconducting state, implying a negative contribution to the free energy, often referred to as the condensation energy. However, because the superconductor is hostile to B-field, when a field B_a is applied as in Figure 4.4B it is forced to divert in order to avoid the bulk of the superconductor. This diversion gives a field energy penalty, an increase of $B_a^2/2\mu_0$ per unit volume of the material. Hence we see that the critical field will be reached when this field energy increase would exceed the condensation energy. These ideas explain the existence of the critical field and also why it must vanish at T_C (since the condensation energy at T_C does also).

Two important points must be made before we leave this section. First, we have talked above about "persistent surface currents". Clearly these currents cannot flow right on the surface, since that would require an infinite current density at the surface. Thus there is a penetration depth (usual symbol λ) into the superconductor describing the length scale in which the currents flow. A typical value is $0.05\,\mu m$ well below T_C, rising to infinity at T_C where the field finally penetrates. We shall return to the calculation of this effect in a later section.

The second point is about sample geometry. Let us consider the magnetisation M of our lump of superconductor. We use the usual notation that

$$B = \mu_0(H+M),$$

where $H(=B_a/\mu_0$ in the above notation) is the applied field and B is the average field over the whole of our lump of superconductor. The Meissner effect indicates that $B=0$, i.e. $M=-H$. This means that the superconductor is an ideal diamagnet with magnetic susceptibility equal to -1. Hence, the discussion so far would suggest that the magnetisation curve would look something like Figure 4.5A, with $M=-H$ up to the critical field H_C and

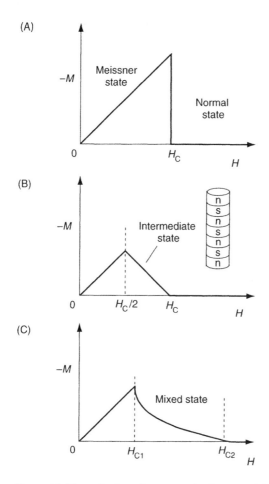

Figure 4.5 Magnetisation of superconducting materials. (A) Ideal magnetisation of a type I material; (B) A transverse cylinder, showing the intermediate state; (C) Ideal magnetisation of a type II material.

then $M=0$ when the sample becomes normal. In practice this is what happens in many superconductors if they have a long needle-shaped geometry with respect to the applied field. For a pure sample, the curve is reversible; there is no history dependence.

However reality is much more diverse. As an example, consider first a cylindrical sample of a pure superconductor, with the field applied transversely. The curve is still reversible, but it now looks as in Figure 4.5B. The point is that the perfect diamagnetism in the so-called Meissner state leads to the field at the edge of the cylinder exceeding H_C when the applied field is only half this value. Hence, between $H_C/2$ and H_C an "intermediate state" is formed in which there are laminae of normal and superconducting material. See Figure 4.5B. A material which forms this type of structure is called a type I superconductor. The scale of the laminae is fine in relation to the diameter of the cylinder, but coarse on an atomic scale; for example a typical laminar thickness is $10\,\mu m$ for a $2\,mm$ diameter cylinder of tin.

This intermediate behaviour is found in extreme form for a sheet of superconductor with the field normal to the sheet. There is no Meissner region since the field has to penetrate at the start. And in a type I material it penetrates in irregular but macroscopic islands.

But you might guess that the statement "type I" leads to the fact that there also exist "type II" superconductors. In these materials, even in the ideal needle-like geometry and with a pure material like Nb, the magnetisation curve looks like Figure 4.5C rather than Figure 4.5A. There are now two critical fields. As the applied field increases through H_{C1}, the material leaves the Meissner state ($M = -H$). But it still has zero resistance and it does not become fully normal until H_{C2} is reached. The region between H_{C1} and H_{C2} is called the "mixed state". The difference between type I and II is in the interfacial energy to form a normal–superconducting (N–S) interface in the material. Such interfaces are energetically un-favoured in type I, whereas they are favoured in type II. Hence, in spite of the energy penalty in distorting the magnetic field, a type I material is reluctant to form interfaces, leading to the fairly coarse scale of the intermediate state in our tin cylinder. However, the situation changes for type II. Here field penetrates forming as many interfaces as possible, in fact forming lines of quantised flux. And this brings us to the third "big idea" to be discussed in the following section.

4.1.3 Flux quantisation

There is a beautiful experiment which has demonstrated the existence of flux lines. It relates to our discussion above about what happens to a sheet of an ideal type II superconductor when a magnetic field is applied normal to it. The sheet is held horizontally in a vertical field, and the emergence of the magnetic field from the top surface of the sheet is examined by sprinkling a fine ferromagnetic powder on it. The powder collects where the field is strongest, and so stains the normal regions of the sheet. A typical result is shown in Figure 4.6A. The spots thus highlight the emergence of flux lines from the material. A flux line consists of a normal core of material which carries the flux, surrounded by superconductor; a circulating current around the core ensures that $B = 0$ in the bulk of the superconductor.

The flux lines are quantised for reasons analogous to the existence of quantised vortex lines in superfluid ^{4}He (Section 2.3). The circulating

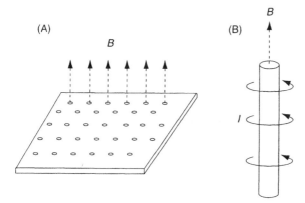

Figure 4.6 Flux lines in a sheet of type II superconductor. (A) Emergence of flux lines from a sheet of superconductor with a field applied along its normal; (B) Anatomy of a flux line – a normal core with circulating supercurrrent.

current around the core has to be provided by superconducting electrons described by a single coherent wave function. And the requirement that the wave function should "eat its tail" on a circuit around the core leads to the minimum magnetic flux allowable being equal to $h/2e$, the so called flux quantum Φ_0 (see Section 4.2.2). Note that the type II material produces as many flux lines as it can (thereby maximising the amount of N–S interfaces) so that all the observed flux lines are singly quantised. Increasing the applied field increases the number of flux lines per unit area of the sheet. The value of the flux quantum can be determined experimentally, simply by counting the number of the lines and knowing the total flux from the applied field and area.

The analogy between quantised vortex lines in superfluid ^{4}He (Section 2.3) and these quantised magnetic flux lines in a superconductor is a productive one. Again the key is to postulate the existence of a large-scale coherent wave function of the form $\Psi = C\exp(i\phi)$ to describe the properties of the superfluid. This same simple expression describes the properties of the superconducting conduction electrons in our metal. Again there is phase coherence in the superconductor on a macroscopic scale, which leads to a lot of exciting consequences, both theoretical and practical.

4.1.4 Summary

In this section we have introduced three basic ideas. First, a superconductor has vanishing resistance, $R=0$. This is analogous to the vanishing viscosity of the slippery superfluid ^{4}He. Second, a superconductor at low applied fields is a perfect diamagnet, $B=0$ in the bulk material. One result of this is that magnetic levitation is possible. However, superconductivity is destroyed by high enough magnetic fields which can limit some practical applications. Third, we have seen that there are large-scale quantum effects, associated with a coherent wave function. A type II superconductor contains quantised flux lines, analogous to the quantised vortex lines in superfluid ^{4}He. We shall now explore these ideas further.

4.2 THE WAVE FUNCTION AND ELECTRODYNAMICS

As in Section 2.3, we shall adopt a simplified postulatory approach to understand what is going on. Again we start with the assertion that we can describe the superfluid condensate by a wave function

$$\Psi = C \exp(i\phi). \tag{4.4}$$

To start with, since we do not know the precise nature of the superconducting state, we suppose that this wave function describes "particles" of mass M and charge Q whose properties are yet to be determined. We shall see that this one postulate explains both the Meissner effect and the existence of flux lines.

First we note that the amplitude of the wave function in equation 4.4 again relates to the strength of the superconducting state. That is, C^2 gives the number density of the particles associated with the superfluid. Again, the phase of the wave function can be reasonably assumed to be given by $\mathbf{p.r}/\hbar$ where $\mathbf{p}$ is the appropriate momentum. But now, unlike in Section 2.3, we must recognise that the particles have charge as well as mass. The effect of this is not too hard to describe, although it does not appear in many undergraduate courses in quantum mechanics. The essential idea is that there are now two terms in this momentum $\mathbf{p}$. One is the familiar $M\mathbf{v}$ term, mass times velocity. But the charge and magnetic field also conspire to change the phase of the wave function, giving another term which contributes to $\mathbf{p}$, namely $Q\mathbf{A}$ where $\mathbf{A}$ is the magnetic vector potential (satisfying $\mathbf{B} = \text{curl } \mathbf{A}$). Thus when we take the gradient of the phase (compare equations 2.2 and 2.3) we now obtain

$$\hbar\nabla\phi = M\mathbf{v}_s + Q\mathbf{A}. \tag{4.5}$$

The requirement that the phase should behave properly now can be used to discuss the consequences for the electromagnetic behaviour of the superconductor. When we consider the circulation around a loop within the superconductor, the phase difference must be zero or a multiple of 2π (compare the discussion in Section 2.3.1).

4.2.1 Bulk superconductors and the Meissner effect

To start with, we restrict our consideration to a simply connected lump of the superconductor. In this case, since we can choose any loop at all, the topological constraints allow only for the phase difference to be zero. This leads to the result

$$M \operatorname{curl} \mathbf{v}_s + Q \operatorname{curl} \mathbf{A} = 0. \tag{4.6}$$

Physically, this equation relates a supercurrent, driven by $\mathbf{v}_s$, to a magnetic field $\mathbf{B}$, related to the vector potential $\mathbf{A}$ by $\mathbf{B} = \operatorname{curl} \mathbf{A}$. In our simple model, we use the usual relation between current density and carrier velocity to give the supercurrent density

$$\mathbf{J}_s = C^2 Q \mathbf{v}_s. \tag{4.7}$$

Equation 4.6 can therefore be rewritten in terms of $\mathbf{B}$ and $\mathbf{J}_s$ as

$$\left(\frac{M}{C^2 Q}\right) \operatorname{curl} \mathbf{J}_s + Q \, \mathbf{B} = 0. \tag{4.8}$$

Equation 4.8 can now be combined with Maxwell's equations to derive the magnetic properties of the superconductor. Because of the infinite conductivity, there can be no DC electric fields inside the superconductor. DC supercurrents are driven only by magnetic fields, and the appropriate Maxwell's equation is simply

$$\operatorname{curl} \mathbf{B} = \mu_0 \mathbf{J}_s. \tag{4.9}$$

We can eliminate either $\mathbf{J}_s$ or $\mathbf{B}$ between these two equations by taking the curl of one of them. In both cases the vanishing of the divergence of either quantity simplifies the vector expansion of curl curl ($= \nabla \times \nabla$) to $-\nabla^2$. To find the properties of the B-field we take the curl of equation 4.9 and substitute curl $\mathbf{J}_s$ from equation 4.8. This gives

$$-\nabla^2 \mathbf{B} + \left(\frac{1}{\lambda^2}\right) \mathbf{B} = 0, \tag{4.10}$$

where $\lambda^2 = M/(C^2 Q^2 \mu_0)$. This equation describes exponential decay. The scale length λ is called the penetration depth, since it gives the distance into a superconductor that the magnetic field penetrates. Equally, it is the length scale in which the surface currents exist. For example in the case of a plane flat N–S interface at $x = 0$, with the applied field in the plane of the interface, the solution of equation 4.10 is simply $B(x) = B(0) \exp(-x/\lambda)$ where $B(0)$ is the field at the surface.

To summarise this discussion so far:

1   We have postulated the existence of a macroscopic wave function, whose phase gradient is related both to Mv momentum and also to QA momentum.

2   We have assumed that Maxwell's equations still apply.

3   These ideas together show that a static B-field can only penetrate a distance of scale λ into the superconductor.

4   Thus $B=0$ in the bulk of the superconductor, i.e. we have seen that the Meissner effect follows from the assumptions.

5   Currents flow only in this surface layer, so that the bulk of the superconductor is a current-free region.

6   Finally we note that the penetration depth varies inversely as C, the wave function amplitude, so that it will diverge as the superconductor is warmed towards T_C.

4.2.2 Superconducting loops and quantised flux lines

Following again the discussion of superfluid ^{4}He in Section 2.3, we now consider a loop of superconducting material, or indeed any lump of the material with a hole through it. When we integrate equation 4.5 around a contour enclosing the hole, the most we can say is that the phase difference must be $2n\pi$ with n an integer. Hence we now have

$$M \oint \mathbf{v_S}.d\ell + Q \oint \mathbf{A}.d\ell = 2n\pi\hbar = nh. \tag{4.11}$$

But if we ensure that the contour does not stray within a penetration depth of the surface, we have just seen that the current $\mathbf{J_S}$ will be zero everywhere along the contour. Therefore the first integral in equation 4.11 must be zero. The second integral has a simple physical meaning, when we recall that $\mathbf{B}=\text{curl } \mathbf{A}$. Stokes theorem tells us that $\oint \mathbf{A}.d\ell$ around the contour equals the flux of curl $\mathbf{A}$, i.e. of $\mathbf{B}$, through the contour. Hence equation 4.11 rearranges as

$$\Phi = n\left(\frac{h}{Q}\right) = n\Phi_0, \tag{4.12}$$

where Φ is the magnetic flux through the loop, and $\Phi_0 = h/Q$ is the "flux quantum" for the superconductor.

This discussion relates to a superconductor with a hole through it. This "hole" can be a physical hole in the metal such as a doughnut shape or a loop. But the same argument applies to an insertion of a thread of normal metal, in which the superconducting wave function amplitude C is zero, thus freeing the phase from its bulk constraints. This exactly follows the arguments for vortex lines in superfluid ^{4}He. So as we have seen in the previous section in a type II superconductor, where formation of N–S interfaces is energetically favoured, application of a B-field causes as many flux lines to form as possible. Hence each flux line carries the minimum flux possible, namely Φ_0.

Therefore, the flux quantum is capable of determination by experiment by at least two methods. One method is simply to count flux lines in a type II superconductor, as mentioned above, and to assume that each line represents a single flux quantum. The other type of experiment is to use a cylindrical tube of superconductor and to cool it through T_C in the presence of a small axial field, which is then switched off. The trapped magnetic flux through the tube is then measured. As expected it is found that the flux does not take any arbitrary value, but it is an integral number of quanta. The superconducting flux quantum so determined is found to be close to 2×10^{-15} Wb, a small number[1] so that the measurements just alluded to are not that easy.

This number is actually of great significance, since it enables the charge of the superconducting "particles" to be determined by experiment. And the answer is $Q = 2e$ where e is the electronic charge, since $h/2e = 2.07 \times 10^{-15}$ Wb. What this shows is that the superconducting state is formed not by single electrons, but by pairs of electrons. Two odds make an even. It is evidence of the way in which a Fermi system conspires to cheat statistics and to imitate Bose–Einstein symmetry. These electron pairs are called Cooper pairs, the basic of the BCS theory of superconductivity, as introduced in Section 1.4.2, and which we shall look at in the next section.

4.3 BCS THEORY AND ITS CONSEQUENCES

When Bardeen, Cooper and Schrieffer (BCS) unveiled their theory in 1957 it gave the answer to several puzzles. One of these was the "isotope effect". Experimental studies in the early 1950s of isotopically pure samples, notably of tin, had shown that there was a dependence of T_C for a specific metal on the isotopic mass M_i, namely $T_C(i) \propto M_i^{-1/2}$. This experiment showed that, in these superconductors at any rate, movement of the ions i.e. phonons must be involved in the mechanism of the superconducting state. Another way of stating the result is that T_C of an isotope is proportional to a typical phonon energy, i.e. to the Debye temperature Θ_D of the lattice.

---

1 At the risk of sounding like an astronomer, a millimetre sized hole in a superconductor when cooled in the Earth's field would trap of order 100,000 quanta.

About the same time, Fröhlich and Bardeen independently started to explore the theoretical possibilities for the involvement of phonons in superconductivity. An important breakthrough was made by Cooper, who pointed out that even a small attractive interaction between electrons in a degenerate Fermi gas could lead to bound pairs of electrons, i.e. to what we now call Cooper pairs.

But the subtle cause of this attractive interaction awaited the BCS theory a few years later. After all, anyone knows that electrons must repel each other with a strong repulsive Coulomb interaction. However, for electrons with energies within about $k_B\Theta_D$ of the Fermi energy, a "second-order" process is possible in which a short-lived phonon is emitted by one electron and is absorbed by a second electron.

Such a process is illustrated in Figure 4.7, in which an electron with wave vector $\mathbf{k}$ exchanges a phonon (wave vector $\mathbf{q}$) with a second electron (wave vector $\mathbf{k}'$). Now the phonon energy, $k_B\Theta_D$, is typically a factor of order 100 to 1000 smaller than the Fermi energy. Thus, when we allow for energy conservation it is clear that only electrons close to the Fermi surface can participate in this type of process. When we remember that momentum (wave vector) must also be conserved, it is a pretty piece of geometry to show that it is much easier to find appropriate states, i.e. the process is much more probable, when $\mathbf{k} = -\mathbf{k}'$.

Hence, when the pair of electrons have opposite momenta, the phonon-mediated interaction causes a small positive attraction which can outweigh the screened Coulomb repulsion between the pair. In hand-waving terms, one electron (moving east-west, say) travels through the metal lattice and in doing so leaves a track in the lattice which arises from the strong attraction between the electrons and ions. This track is then sensed by another electron travelling west-east. The fact that the electrons are moving fast in opposite directions means that their Coulomb repulsion is only effective

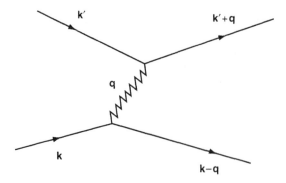

Figure 4.7 A second-order electron–electron interaction involving emission and reabsorbtion of a virtual phonon, wave vector **q**.

for a short period of their life, whereas the phonon track, takes a long time to disperse so that the attraction has a long range. So it turns out that, for strong enough electron–phonon interaction, the nett electron–electron interaction is attractive, and hence superconductivity occurs.

4.3.1 The superconducting ground state and the transition temperature

The result of forming these Cooper pairs is quite dramatic. In the super-conducting ground state (at $T=0$), the BCS picture is that all the electrons conspire to occupy (or not to occupy) states in pairs with opposite momenta and opposite spins. In an obvious shorthand, the occupations of the electron states $+\mathbf{k}\uparrow$ and $-\mathbf{k}\downarrow$ are correlated. We say that each Cooper pair has total spin $S=0$ and total momentum $L=0$. For this correlated state of the whole system, it is found that the energy of the ground state is reduced below that of the normal metal. Hence the new superconducting state. It is important to recognise that this is a collective state in which all the Cooper pairs play a part and in which they are constantly scattering into one another. After all, a single Cooper pair is a strange animal, in that the two components are travelling fast in opposite directions.

The magnitude of the effect clearly depends on (i) the strength of the electron–phonon interaction, since this is the mechanism of the attractive interaction, (ii) the density of electron states around the Fermi level, since this governs how many electrons can feed from this attraction and (iii) the Debye temperature, since that is the phonon energy scale. In order to get a result, BCS used a simplified model in which all anisotropy was removed, leaving a spherical Fermi surface with density of states $N(0)$.[2] The electron–phonon interaction is modelled by a single constant matrix element V between electrons within $k_B\Theta_D$ of the Fermi surface and zero otherwise. Their calculation for the transition temperature gives

$$T_C = 1.14\,\Theta_D \exp\left(-\frac{1}{N(0)V}\right). \tag{4.13}$$

It is worth remarking that any a priori calculation of T_C is notoriously difficult, since it is always the case that $T_C \ll \Theta_D$. Hence any minute variation in V or in $N(0)$ gives a very large change in T_C calculated from equation 4.13.

The first achievement of the theory is to give a reasonable picture of the superconducting ground state which is consistent with the postulates

---

2 We use notation $N(0)$ for the density of states at the Fermi level, as is usual in this field when the excitation picture is used. The quantity $g(E_F)$ in the vacuum picture, more commonly used in general statistical physics, refers to exactly the same quantity.

made earlier in this chapter. It gives a ground state involving all the electrons which is a single coherent whole, so that it can indeed be described by a single phase-coherent wave function.

4.3.2 Excited states and the energy gap

The other great advance of the BCS theory is that it gives a picture of the excited states, as well as one of the ground state. The essential point is that to create an excited electron from the ground state, we must eject a Cooper pair, thereby making a pair of excited particles. And this requires us to break all the attractive correlations which the pair enjoyed in the superconducting ground state. There is thus a minimum energy per electron to break free, usually written as $\Delta(0)$ at $T=0$. This energy gap is again a measure of the strength of the superconducting state, and BCS showed it to be directly proportional to T_C. In fact the relation for the simplified BCS model is that the minimum energy to break a pair is $2\Delta(0)=3.52\,k_B T_C$.

We noted in Section 1.5 the importance of the excitation dispersion relation to superfluidity, so let us consider the influence of an energy gap in this regard. First, a word of caution. The superconducting ground state, as we have seen, is a many-body concept, so it is rather odd to be projecting it on to a one-particle picture. The most useful way of attempting this is to work in the "excitation" picture, in which the energy zero is taken to be at the ground state as illustrated in Chapter 1 (Figure 1.7). Let us briefly review these ideas, as related to identical Fermi particles in a box, the simplest picture of electrons in a metal. There are three steps in the argument, corresponding to the three parts of Figure 1.7.

1. *Vacuum picture*   The usual one-particle picture is to start with a vacuum and to enumerate the quantum states. For the free particles this gives the simple parabolic dispersion relation

$$E_k = \frac{\hbar^2 k^2}{2m},\tag{4.14}$$

where m is the electron mass. [Note that in metal physics it is conventional to work with wave vector $\mathbf{k}$ rather than with momentum p.] In a real metal, the detailed form of this expression will be considerably modified by the effective lattice potential experienced by the electrons, but this does not affect the central idea here. In the ground state, electron states are filled up to the Fermi level E_F and are empty above it. Excitations are formed in pairs by taking an electron below E_F and promoting it to a previously unfilled state above E_F, leaving in a hole below in the "Fermi sea". There is a very large ground state energy in this picture, arising from the Pauli exclusion principle.

2. *Excitation picture – normal metal* The vacuum picture gives a good view of the ground state, and in particular of the Fermi surface and E_F. However, for consideration of the excitations, it is better to use a new picture which recognises that there is energy input associated with the creation of both the hole and the excited electron. This picture takes $E=0$ at the ground state, and simply plots the $E–k$ relation for excitations. Excitations are hole-like if $k < k_F$ and particle-like if $k > k_F$, and all energies are positive. For the free electron gas, the dispersion becomes

$$E_k = |\hbar^2 k^2/2m - \hbar^2 k_F^2/2m| \equiv \varepsilon_k, \tag{4.15}$$

defining the (normal) excitation energy ε_k. We may note that this excitation energy has a minimum value of zero, corresponding to excitations formed virtually at the Fermi level.

3. *Excitation picture – superconductor* The meaning of the energy gap is that, even for excitations at $k=k_F$, a finite energy is now needed for their creation. The BCS result is simple, namely that

$$E_k = + \surd(\varepsilon_k^2 + \Delta^2). \tag{4.16}$$

Note that this means that at $k=k_F$ we have $E=\Delta$, whereas well away from k_F there is no change from the normal state, i.e. $E=\varepsilon$.

The dispersion relations for normal and superconducting metals of equations 4.15 and 4.16 are sketched in Figure 4.8A. It is worth remarking that Δ is a very small energy compared to $E_F(\Delta/E_F$ is typically 10^{-5} only) so that there is little asymmetry between particle and hole branches. Another aspect of note is that this energy gap is quite unlike the usual energy gaps in condensed matter physics which give rise to insulating or semiconducting behaviour. These derive from effective lattice potentials, and are (roughly!) independent of the electronic properties. In contrast, by its origin, the superconducting gap is tied to the Fermi level itself. Furthermore, it relates to excited states and not to the ground state (which of course is a good – a super – conductor) in spite of the "Fermi level lying in the gap".

The dispersion curves are intimately associated with the density of states of the excitations. The **k** values of electron states are determined by geometrical requirements of fitting waves into boxes. For the normal metal in the vacuum picture, the uniform distribution of states in **k** combines with the smooth dispersion relation of equation 4.14 to give the well-known relation that the number of states per unit energy range is proportional to $E^{1/2}$. In the superconductor, the **k** values of the states are unchanged, but the energy gap and corresponding flat portion of

(A)

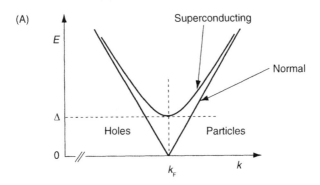

(B)

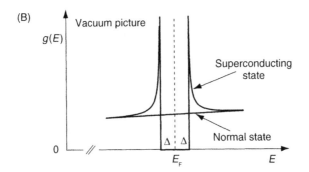

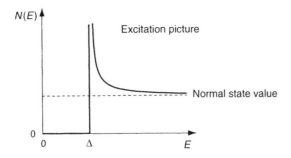

Figure 4.8 (A) Dispersion relations in the excitation picture for a normal and a superconducting metal. (B) Density of states of a BCS superconductor.

equation 4.16 leads to a hole in the density of states at energies less than $\pm\Delta$ from E_F and a corresponding piling up of the states at energies just outside this, as illustrated in Figure 4.8B. Note that this illustration greatly exaggerates the magnitude of the gap in the vacuum picture, so that the normal density of states $N(0)$ in the excitation picture can be considered a constant, independent of E. The horizontal part of the dispersion curve of equation 4.16 generates a weak infinity in the (energy) density of states at $E = \Delta$, with a dependence $N(0)E/(E^2 - \Delta^2)^{1/2}$ when $E > \Delta$.

4.3.3 Some consequences of the energy gap

The existence of the energy gap has profound consequences for the observable properties of the superfluid, some of which are discussed in the next section. For example, all the equilibrium properties at low temperatures are dominated by terms involving $\exp(-\Delta/k_B T)$, often referred to as a "gap Boltzmann factor" although actually one is applying Fermi not Boltzmann statistics to the excitation gas. This dependence dominates the electronic thermal conductivity and heat capacity well below T_C.

Another consequence, following Landau's arguments outlined in Chapter 1, is the existence of a critical velocity, equal to $\Delta/\hbar k_F$ as one can see from the dispersion curve in the vacuum picture. This is what makes the superconductor a superfluid. We saw in ^{4}He that the Landau velocity itself was only observed directly in a few types of experiment, notably in ion motion. Usually flow experiments are limited by vortex production. Similarly in a superconductor, practical critical currents are modified by the existence of flux lines, themselves an inevitable result of currents caused by the flow of charged particles.

So far our discussion of the gap has referred to $T = 0$. We have seen that $\Delta(0) = 1.76\,T_C$, i.e. the gap and the transition temperature have the same energy scale. Hence, as we have just noted, as soon as the temperature rises to a significant fraction of T_C, there will be a substantial number of single particle excitations, i.e. of normal fluid in the language of Chapter 2, given by the gap Boltzmann expression. But there is a cooperative effect here. The excitations themselves are formed by breaking Cooper pairs, hence weakening the ground state. So when the number of excitations becomes significant, the strength of the superconducting state must become weaker, thus making the energy gap smaller. But the smaller gap means yet more excitations than the "number you first thought of" – there is a self-consistent procedure to be followed here, tractable by numerical solution. The result, shown in Figure 4.9, is that the gap collapses to become zero at T_C.

The idea of the energy gap allows us to estimate the condensation energy, the energy gain when the superconducting state is formed at $T = 0$. We have seen in Section 4.1.2 above that the condensation energy balances the field energy at the critical field, i.e. it equals $\mu_0 H_C^2/2$ per unit volume of superconductor. An estimate of the condensation energy is visible by inspection of the density of states graph of Figure 4.8. The

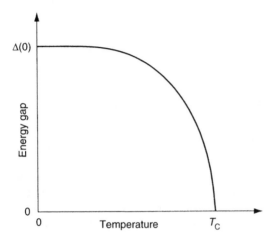

Figure 4.9 The BCS superconducting energy gap.

difference in energy between the normal and superconducting states is the same as an energy of order Δ supplied to $N(0)\Delta$ electrons, so that the condensation energy at $T=0$ is of order $N(0)\Delta(0)^2$. Comparison with the above shows that we expect $H_C(0)$ therefore to be proportional to $\Delta(0)$. This is a nice result, since we have now seen that both T_C and $H_C(0)$ are proportional to the same $\Delta(0)$ in BCS theory. So we have understood why there is a law of corresponding states in BCS theory, a single scaling parameter on both axes of the B–T phase diagram (Figure 4.3).

Finally in this section, it is interesting to consider the nature of the excitations implied by the dispersion relation, equation 4.16. Looking at the dispersion curve in the excitation picture (Figure 4.8A), it is evident that in the normal metal these excitations change in character discontinuously at k_F. For $k>k_F$ the excitations are particles, with group velocity $\partial E/\hbar \partial k$ equal to $+v_F$, the Fermi velocity. For $k<k_F$ they are holes, with group velocity $-v_F$ in the opposite direction to their momentum. Normal state excitations have unambiguous particle or hole character. However, the superconducting state is quite different. Here, the group velocity is zero very close to k_F, and only reaches $\pm v_F$ when $E \gg \Delta$. At energies close to Δ the excitations have mixed particle-hole character. The full BCS treatment makes this quite clear. The nature of an excited state is that it is a coherent mixture of particle and hole, hardly surprising when one recalls that it is formed from the highly correlated ground state. This opens the lid on quite a large topic, since it turns out that this coherence has an important effect on many of the observed properties of the superconductor.

4.4 SOME OTHER PROPERTIES OF SUPERCONDUCTORS

We shall now discuss further some of the observed properties of super-conductors, with particular emphasis on how these are understood in terms of the BCS model and the coherent wave function. It must be remembered that BCS theory is based on the simplest approximation of the electronic structure, neglecting all influence of band structure.

4.4.1 Determination of the energy gap

The value of the energy gap may be performed by at least three types of measurement. We exclude a theoretically possible method based on Landau critical velocity, for reasons of the confusing secondary effects arising from flux lines. The results on real superconductors confirm that the BCS result $\Delta(0) = 1.76\ k_B T_C$ is a fair approximation, although not an exact result. This is hardly surprising, since it is based on a simplified ideal model of the electronic structure, the phonon structure and the electron–phonon interaction of the metal.

1. Thermodynamic measurements The internal energy of the electrons can readily be calculated from the usual expression for a Fermi gas $U = \int Eg(E)f(E)\,dE$ where $g(E)$ is the density of states and $f(E)$ the Fermi–Dirac distribution function. Given the density of states of the excitations, together with the dependence of the gap on temperature, we can thus compute the internal energy and hence the electronic heat capacity C_e of a BCS superconductor. Since superconductivity is a low temperature phenomenon, the contribution of the lattice to the heat capacity is usually small, or at any rate known well enough to be subtracted from the total heat capacity. At low temperatures, typically below about $0.3T_C$, the expression for C_e is dominated by the gap Boltzmann factor $\exp(-\Delta(0)/k_B T)$. Hence $\Delta(0)$ can be extracted from the experimental data.

An alternative is to measure the electronic contribution to the thermal conductivity K_e when ordinary impurity dominates. From the kinetic theory expression $K_e = 1/3 C_e v_F \ell$, where ℓ is the mean free path for excitations, effectively a constant. Hence again the low temperature behaviour yields $\Delta(0)$ via the gap Boltzmann dependence. [As a remark aside, this method is not suitable for Pb and Hg, where the ratio T_C/Θ_D is becoming uncomfortably high (0.08 for Pb, 0.04 for Hg, compare 0.003 for Al). This is for two reasons. One is that BCS theory only works well for so-called weak coupling materials which have small values of this ratio. The other is that, in these pure metals, interpretation is confused by electron–phonon scattering for which the theory is less secure.]

2. Spectroscopic methods "It works in a semiconductor, so why not in a superconductor?" is a good question. In a semiconductor there is an optical absorbtion edge when the optical photon energy passes through the band gap. Similarly here, one expects a very well defined pair-breaking edge when the quantum of energy passes through 2Δ. When we put in the numbers for a superconductor with a T_C of 1 K, we find that the feature should occur at $T=0$ at a frequency of $3.52\,k_BT_C/h=70\,\text{GHz}$. For photons, this is in the microwave region with a wavelength of about 4 mm. In practice therefore the experiments are done in a cavity at fixed frequency as a function of temperature, probing the dependence of the gap with temperature. Without going into detail, it all fits the picture very well, with (for example) very low losses at a low enough temperature when a cavity frequency less than $2\Delta/h$ is employed. It is also possible to use ultrasonic phonons, although here the typical frequency required to probe the gap at $T=0$ is uncomfortably high for conventional methods. However, these can be used to explore the gap near T_C.

3. Tunnelling experiments The most direct and clean experiments to study the energy gap are those involving electron (quasiparticle) tunnelling through an insulating barrier. The tunnel junction is typically an oxide barrier, which forms the only connection between two evaporated metal strips. The measurement made is of the current–voltage (*I–V*) relation between the two strips. When the metals are normal, Ohm's law applies with *I* simply proportional to *V*. Naturally the magnitude of the tunnelling resistance depends (linearly) on the area of the contact of the junction and (pathologically, exponentially) on the barrier thickness, since it depends on the electron wave function tunnelling across the forbidden insulating region. It is therefore a measurement of little fundamental interest, even though it theoretically probes the electron density of states in the two metals.

With one or both metals superconducting, the measurement suddenly becomes interesting! This arises from (i) the violent gap structure (Figure 4.8) injected into the density of states of excited electrons and (ii) the existence of the Cooper pairs which make up the superconducting condensate. Tunnelling of the condensate wave function does occur for very thin barriers, of order 1 nm only. This is the story of the Josephson effects, to be discussed in a later section. We concern ourselves here with tunnelling of the excitations, in practice the only effect for slightly thicker barriers, typically 5–10 nm.

The main features of the *I–V* characteristic can be understood in relation to the density of states picture of Figure 4.8. For simplicity we consider as a representative experiment, a superconductor–insulator–superconductor (or SIS) junction, in which both metals are the same superconducting metal. When a voltage *V* is applied between the two

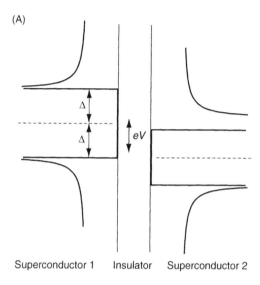

Superconductor 1   Insulator   Superconductor 2

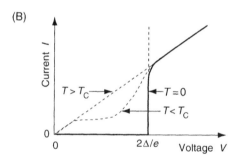

Figure 4.10 Behaviour of an SIS tunnel junction. (A) Density of states picture of an SIS junction; (B) Typical current–voltage characteristics.

metals, virtually all the potential drop is across the insulator. Tunnelling occurs at a constant energy only, so that the time-dependence of the wave functions match on the two sides, allowing the weak leakage between the two sides to have significance. Hence the densities of states as described by the vacuum picture are represented as in Figure 4.10A.

At $T=0$, bearing in mind that the Fermi level lies at the centre of the 2Δ energy gap, the lower states are all filled and the upper ones are empty. Thus, for small applied voltages as illustrated, there are no horizontal tunnelling processes possible so that there is no tunnelling current. This situation persists until the condition $eV=2\Delta$ is reached, at which the piled up filled states on the left line up with the piled up empty states on the right, and there is a rapid current onset as illustrated in Figure 4.10B. It is worth pointing out here that the tunnelling process generates two excitations, a hole on the left and a particle on the right; another way of looking at the process in the excitation picture is that a Cooper pair on the left is broken (using the 2Δ energy available) to form these two excitations, one of which tunnels through the barrier.

At higher temperatures, there are some thermal excitations which allow some tunnelling to occur of particles (left to right) and of holes (right to left) below the gap threshold, leading to the type of I–V curve also shown in Figure 4.10B. These characteristics all give detailed information about the density of states, and provide excellent confirmation of the broad validity of the BCS approach in many materials. Of course, to be exact, any theory must include the real electron–phonon interaction and the real electronic structure of the metal in question. For instance, it is interesting that the tunnelling characteristic above the gap shows subtle features which reflect the phonon spectrum of the metal.

4.4.2 The phase transition

So far we have understandably concentrated more on the properties of the superconducting state than on the transition to superconductivity. However, the transition itself turns out to be an ideal textbook affair, providing an almost unique example of an ideal *second-order transition*.

Let us first look at the experimental signature of such a transition, in conventional thermodynamic language. A first-order phase transition is one in which the first derivatives of the free energy (such as entropy and volume) are discontinuous at the transition point. Examples include the usual gas–liquid, gas–solid and liquid–solid transitions as mentioned earlier in Chapter 1. The relation between the discontinuities and the behaviour of the P–T phase diagram is enshrined in the Clausius–Clapeyron equation discussed in Chapter 1. The transition also frequently displays hysteretic supercooling or superheating effects arising from the requirements to nucleate the new phase.

In a classic second-order transition, the changes are more subtle. The entropy and the volume of the two phases are continuous across the transition, but their derivatives (second derivatives of the free energy) are not. So, for example, there is a jump in the electronic specific heat capacity ($C = T dS/dT$) at the transition. This is precisely what is seen in a superconductor in zero applied magnetic field, as shown schematically in Figure 4.11. The entropy curve shows a decrease below T_C from the expected extrapolation of the normal state entropy; and there is a corresponding upward specific heat jump (ΔC) below T_C. One of the successes of BCS theory was the calculation of the magnitude of this jump, within the spirit of the law of corresponding states. As noted above,

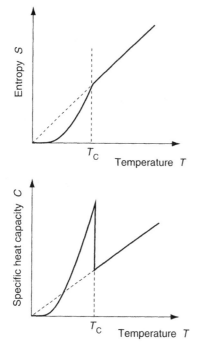

Figure 4.11 Electronic contributions to thermal properties of a BCS superconductor.

the electronic contributions to both C and S well below T_C go rapidly to zero, dominated by the gap Boltzmann factor, $\exp(-\Delta/k_B T)$.

A characteristic feature of a second-order transition is that it is a type of order–disorder transition, in which a new state described by an *order parameter* which is zero above the transition temperature but which appears rapidly below it. Transitions of this type include (i) ferromagnetism based on a mean field theory, (ii) some transitions in alloys, (iii) superfluidity in ^{4}He. In these cases the order parameter is (i) the spontaneous magnetisation, (ii) the fraction of ordered alloy and (iii) the superfluid density, or equivalently the superfluid wave function amplitude squared. In a mean field approach (i.e. an approach which treats long-range order as being the only important order parameter), the phase transition can be understood in terms of an expansion of the free energy in a power series in the order parameter, an approach developed by Landau, and one which leads to an ideal second-order transition. However, in all of these three cases, the mean field approach is seriously flawed in detail, since interactions and hence short-range effects are also important. For this reason all three transitions in practice show

λ-behaviour in the specific heat, the transition being marked by C becoming (weakly) infinite rather than having a finite jump at T_C.

Superconductivity is notable in that the simple long-range order approach is valid. The superconducting state is formed by the subtle phonon mediated electron–electron interaction which is by its nature both weak and long range. The order parameter is again C^2, the amplitude squared of the wave function or equivalently the superfluid density (in the language of the two-fluid model). The theory of the transition and of the behaviour at temperatures a little below it has been developed with great success by Ginzburg and Landau, using the power series approach mentioned above.

The discussion in this section has so far related to the transition in zero magnetic field. Let us finally note that as soon as a magnetic field is applied, the transition becomes first-order rather than second-order. The transition is accompanied by the Meissner effect, involving a complete and sudden change in the magnetic flux distribution together with the sudden appearance of the condensation energy. But also, as implied by the entropy curve of Figure 4.11, these changes are associated with a discontinuous entropy change, that is by a latent heat. This means that, if we adiabatically increase the magnetic field to a superconductor below T_C, then heat will be absorbed when the transition takes place and the superconductor must cool as it enters the normal state.

4.4.3 Length scales in superconductors and N–S interfaces

In Section 4.1.2, near the start of this chapter, we introduced the idea of type I and type II superconductors, in relation to their behaviour in an applied magnetic field. Roughly speaking, a type II material "likes" to develop normal–superconducting (N–S) interfaces, whereas a type I material does not. Hence the type II material welcomes quantised flux lines, whereas type I does not, leading to the different magnetisation curves illustrated earlier in Figure 4.5. We are now in a position to understand these differences more fully in terms of the various length scales which are important in a superconductor.

1. Mean free path The first length scale is that of the electron mean free path ℓ in the normal metal. Leaving aside ultra-pure and strongly coupled superconductors, electron scattering in the temperature region of interest is usually simply by potential (elastic) scattering by impurities, rather than scattering by phonons. These scattering centres will also scatter excitations in the superconducting metal, characterised by the same mean free path ℓ.

As an aside, we may remark that magnetic impurities play a special role in superconductivity and are part of another story. As well as normal elastic potential scattering, processes also occur in which spin-flip inelastic scattering occurs. Such scattering is bound to disturb the integrity of the Cooper pairs, since the states $+\mathbf{k}\uparrow$ and $-\mathbf{k}\downarrow$ are scattered so differently. As a result, such impurities are highly destructive to the superconducting state. As an example, very small amounts of Mn impurity in Al are found to lower the measured values of T_C considerably. However, let us return to the question of length scales.

2. *Coherence length*   The second length is one not yet mentioned explicitly in this chapter, called the coherence length (usual symbol Greek "xi", written ξ). The idea is an important one, and is essentially "how big is a Cooper pair?". We have seen that a Cooper pair consists of two components, given by the electron states $+\mathbf{k}\uparrow$ and $-\mathbf{k}\downarrow$. These represent electrons moving at the Fermi velocity v_F in opposite directions. Clearly this implies some significant length scale for the two halves of the pair to "know about each other". In a pure metal at low temperatures, the obvious scale length is the velocity v_F multiplied by a time which is long enough for energies to be resolved on a scale of $\Delta(0)$. This gives

$$\xi_0 = \frac{a\hbar v_F}{k_B T_C} \qquad\qquad (4.17)$$

where a is a numerical constant.

This type of idea was first suggested by Pippard to help the understanding of microwave surface impedance measurements, through which it became clear that the electrodynamics required a non-local theory. The fields effective at a specific point in the superconductor were considered as those averaged over a region surrounding the point with an exponential weighting factor of $\exp(-r/\xi_0)$. Experimentally for a number of superconductors the value of a was found to be around 0.15, and BCS verified the whole idea, calculating $a=0.18$, another success for their theory. Typical values of ξ_0 for pure metals are 40 nm (Nb), 230 nm (Sn) and 1600 nm (Al).

In practice, as we shall shortly see, there is much interest in superconductors which are not ideally pure, but in which the mean free path is also small on this scale. In this case, the two halves of the pair do not get a chance to travel as far as ξ_0 before they are scattered, so it is a reasonable assumption that the effective coherence length becomes ξ given by

$$\frac{1}{\xi} = \frac{1}{\xi_0} + \frac{1}{\ell}. \qquad\qquad (4.18)$$

A related way of thinking about a coherence length is that it represents the shortest scale over which the wave function can change. There is a certain rigidity to the wave function, since rapid changes in it imply high momentum and hence high kinetic energy. Thus for example it represents the core radius for a flux line, a point to which we shall return.

3. Penetration depth The third important length is the penetration depth, describing the exclusion of magnetic flux from the bulk of a superconductor. As we saw in Section 4.2.1 it is also the depth from a surface in which currents flow. In that section, we used local electrodynamics to derive an expression for the penetration depth λ given after equation 4.10. This was first done by London and the depth $= \sqrt{(M/(C^2 Q^2 \mu_0)}$ is properly referred to as the London penetration depth λ_L.

This earlier result is fully correct if, but only if, the coherence length is very small. The assumption in Section 4.2.1 was that the wave function amplitude was constant, equal to C^2, right up to the surface, and zero by implication outside (in vacuum or in normal region). This assumption therefore presupposes a sharp step in C at the surface, and a sharp step implies a zero coherence length.

In practice, therefore, the influence of the coherence length just introduced is important here. Pippard saw that a non-local electrodynamics would affect field penetration, and that the effective penetration depth would be increased above the London expression for superconductors in which ξ was larger than λ_L. He found that in this limit the effective penetration depth should be increased to $\lambda = (\xi \lambda_L^2)^{1/3}$. In what follows, we shall use the symbol λ for the effective penetration depth, in whatever regime we are considering.

The relative magnitude of λ and ξ turns out to be the prime factor in determining the interfacial energy of an N–S interface. In particular, whether a superconductor is designated as type I or type II depends on the ratio λ/ξ usually written as the single symbol κ.

Let us now see how the magnitude of κ affects the energetics of an N–S interface. Consider a plane boundary between N and S regions under the application of a magnetic field. The boundary is sketched in Figure 4.12 for two situations. The figure shows the expected variation of (i) the magnetic field B near the interface, whose spatial variation is governed by a scale length λ and (ii) the superconducting wave function $|\Psi|$ or C near the interface, whose spatial variation is governed by ξ. There are two contributions to the interfacial energy. First, the penetration of the magnetic field into the superconductor lessens the field energy penalty of the Meissner effect. This means that larger values of λ will favour the formation of interfaces. On the other hand, the rigidity of the wave function excludes a region of the superconductor from the benefit of the superconducting condensation energy. This means that large values of ξ will disfavour the formation of interfaces.

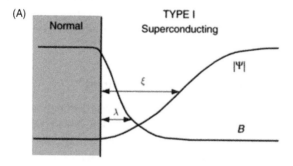

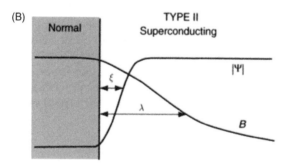

Figure 4.12 Length scales in type I and type II materials, showing the importance of the scale lengths λ for field penetration and ξ for wave function stiffness.

Figure 4.12A illustrates the situation for a small κ material, a type I superconductor. In this case, λ is small giving little advantage from field penetration, whereas on the same scale ξ is large, implying a large energy disadvantage from loss of condensation energy. Hence the interfacial energy is positive and interfaces only in form with reluctance in a type I material.

The reverse is true for a high κ, type II material. Here, there is little loss of condensation energy because of the small ξ, whereas the field is allowed well into the bulk of the superconductor because of the large λ. The interfacial energy is now negative, and hence as soon as a significant field is present, the flux enters as individual singly quantised flux lines.

Detailed calculations of these energies show that the interfacial energy of an N–S interface changes sign at $\kappa=1/\sqrt{2}$, so that this value marks the official changeover from type I to type II. We may note also that the influence of the mean free path ℓ is important in deciding whether a material is type I or type II. Increased impurity or defects in the material inevitably decrease ξ below its pure value ξ_0, with $\xi=\ell$ in the extreme case of small mean free path. Hence, although only Nb and V among the pure metals show type II behaviour, it is much more common for alloys to be type II materials.

4.4.4   Type II superconductors and flux lines

We can now understand the reasons for the magnetisation behaviour of an ideal type II material, illustrated above in Figure 4.5C and repeated in Figure 4.13 This illustrates the magnetisation curve when there are no complications from (i) irreversible effects caused by excessive defects in the material or (ii) demagnetisation factors caused by the shape of the sample of material under study. In other words, we have a long thin pure sample with field applied along the length of the sample.

There are three critical fields shown in Figure 4.13. First there is the bulk or thermodynamic critical field H_C, defined such that the condensation energy per unit volume is $1/2\mu_0 H_C^2$. This is of course valid for any super-conductor, type I or type II. Then there is the lower critical field, defined as the field at which flux lines first penetrate the type II material.

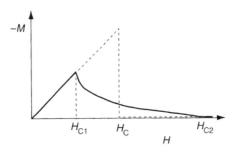

Magnetisation of a reversible type II superconductor

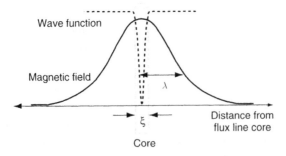

Figure 4.13 Properties near a flux line.

Finally, we have the upper critical field H_{C2}, above which the material is fully normal.

The structure of a flux line is an interesting topic. The flux line is a cylindrical object with two radii. It has a normal core of (scale) radius ξ, with the wave function being zero at the centre in order to allow the phase of the wave function to wind up by 2π around the flux line. But then it has a second (larger – type II!) radius of scale λ marking the range of the magnetic field near the line and the associated supercurrents around it.

When the field is wound up from zero, at first there is a Meissner region, where the condensation energy gained by the sample remaining fully superconducting ensures that it is not worth while for flux to penetrate. But when flux penetration first takes place, only an area of order ξ^2 per flux line loses condensation energy, whereas the lines can lie as much as λ apart whilst yet allowing almost full flux penetration. Hence at H_{C1}, since only a fraction of order ξ^2/λ^2 of the material needs to lose condensation energy, we have $1/2\mu_0 H_{C1}^2 \simeq 1/2\mu_0 H_C^2 \, (\xi^2/\lambda^2)$ and hence $H_{C1} \simeq (\xi/\lambda)H_C = H_C/\kappa$.

Between H_{C1} and H_{C2}, the density of flux lines increases, and the flux line spacing and the fraction of normal material increase with it. Finally at H_{C2}, the flux line cores touch, i.e. the material becomes fully normal, and there is nowhere for the supercurrents to flow. This occurs when $H_{C2} \simeq H_{C1}$ (λ^2/ξ^2), so that $H_{C2} \simeq \kappa H_C$. [A proper calculation gives $H_{C2} = \sqrt{2}\kappa H_C$, which is how the $1/\sqrt{2}$ factor mentioned above enters the scene.]

These are results which are useful, as well as interesting. They hold the key as to how one can start to engineer high field magnets using superconductors. The first requirement is to manufacture a material which has a very high H_{C2} and also reasonably high T_C. This can be achieved by introducing sufficient impurities and physical defects to a material that its ξ value becomes very small. In practice, in conservative technology, a common material is an alloy such as 70 Nb/30 Ti, whose transition occurs at 9 K or so, and whose upper critical field exceeds 7 T. However, this is only a first requirement towards a solution to the design. The next is to realise that an ideal reversible material, such as discussed above, is of no use at all for the purpose. The problem is that a magnet winding must at the same time carry a (super)current and also be bathed in high magnetic fields. Hence a flux line in this situation will experience a large transverse force from the $I \times B$ Lorentz force. If the flux is free to move it will do so, causing leakage of the field and usually heating and thus failure of the superconducting wire. Therefore the trick is to have the wire supplied with many inhomogeneities which can act as pinning centres for flux lines, a goal again achieved by having a lot of defects. Physical defects pin dislocations to make an alloy strong – and they also provide potential barriers over which any flux line has to pass. But that is still not the end of the story. The NbTi alloy magnet wire is usually fabricated by embedding many (e.g. 61) fine filaments of superconductor in a matrix of pure copper. The clever idea here is that if a "flux jump" does occur, it will only happen in one filament, and the current can be shunted via the pure copper through the other filaments, thus making the wire much more stable against catastrophic failure. This is a vital feature since these magnets have enormous amounts of stored energy, which cause great excitement if suddenly let loose in a bath of liquid helium!

Before we leave this topic it is instructive to recall the discussion of vortices in superfluid ^{4}He mentioned in Section 2.3.3. Again we saw that a vortex consists of a core surrounded by a low field. In the case of ^{4}He the core radius is very small, of order 0.1 nm, demonstrating that the coherence length ξ of the wave function is very small. On the other hand, the flow field has a very large "penetration depth", since there is no screening in this case. We saw that the velocity field falls off as $1/r$ and is thus of much larger range than any exponential. Thus, in superconductor parlance, superfluid ^{4}He is an extreme type II material with effectively infinite κ. This is consistent with the very ready formation of vortices in the superfluid.

4.4.5 N–S interfaces and Andreev processes

Another amusing feature of having an N–S interface in any sort of superconductor is the possibility of what is called Andreev reflection of excitations. Since, well after its discovery in conventional superconductors, this phenomenon has turned out to be a major feature in our understanding of unconventional superconductors and of superfluid ^{3}He (Chapter 5), we shall discuss it here.

We use the excitation picture for the excitations. The question is as follows. Consider an electron excitation in a pure normal material, travelling towards the superconductor. Let us suppose that the mean free path is long, so that the motion is effectively ballistic. What happens at the interface? The question is particularly interesting if the excitation has energy less than the energy gap Δ of the superconductor.

The situation is sketched in Figure 4.14. As we have seen, the superconducting state builds up gradually as a function of distance from the interface, with the characteristic length ξ. In the graph we have sketched the expected form of the excitation spectrum at various positions. In the normal state there is no gap, and the gap builds up gradually as we enter the superconductor. So our below-gap excitation has a problem. It is travelling ballistically at constant energy, so what is clear is that in the bulk of the superconductor there are no available states for it at the right energy. Hence, the electron gradually slows down until it reaches point

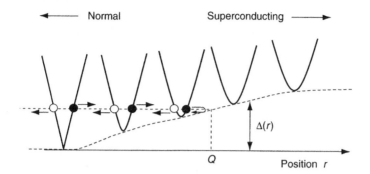

Figure 4.14 Andreev reflection. The diagram indicates the dispersion relation of excitations as a function of position near an N–S interface. A particle with energy less than the full energy gap Δ incident from the normal metal has to be retroreflected as a hole.

Q where it has zero group velocity. What happens then is that it emerges as a hole which travels back into the normal material. To complete the story, keeping track of the particle number requires that a Cooper pair is injected into the superconductor. [One particle on the left generates a hole on the left plus two particles, the Cooper pair, on the right, conserving particle number.]

The process becomes even more striking in three dimensions. Suppose that the incoming electron quasiparticle is in state $\mathbf{k}\uparrow$ with energy E (above the Fermi level – remember that we are working in the excitation picture). The outgoing hole quasiparticle has the same energy and *almost* the same $\mathbf{k}\uparrow$ state (bearing in mind that $E \ll E_F$), but its group velocity is reversed. Thus we have an almost pure retroreflection mechanism. The excitation almost precisely retraces its steps.

In the past, the influence of these effects have been studied on the transport properties of a pure type I superconductor in the intermediate state. A cylindrical sample in a transverse magnetic field (compare Figure 4.5B) enters an intermediate state dominated by alternate normal and superconducting laminae, so that the properties (electrical resistance, thermal resistance and even the thermoelectric power[3]) see the influence of the N–S interfaces. There is a considerable boundary thermal resistance, since the excitation (and its energy) is reflected back at an interface; but with only a small effect on electrical resistance since current through the boundary is virtually unchanged because of the injection of the

3 One of the author's past vices!

Cooper pair into the superconductor. Measurements are also made nowadays using nanostructure fabricated devices, and the Andreev reflection properties are quite a tool in exploring the nature of the superconducting state in unconventional superconductors.

4.4.6 Josephson effects

In our earlier discussion of electron tunnelling, we mentioned the Josephson effects. These occur when two superconductors are connected by a "weak link", meaning a weakly superconducting link. In practice weak links can be provided by very thin (about 1 nm) oxide layers, by delicate point contacts, by fabricated microbridges, or by other nanofabrication techniques. Josephson tunnelling then occurs in addition to the quasiparticle (excitation) tunnelling discussed earlier. It is tunnelling of the Cooper pairs themselves, and it arises from the coupling of the superconducting wave function through the weak link.

Josephson suggested two basic equations to describe how the difference in phase of the superconducting wave function between the two sides of the weak link would be determined by external conditions. Suppose that the two sides have wave functions given by $\Psi_1 = C_1 \exp i\phi_1$ and $\Psi_2 = C_2 \exp i\phi_2$. The Josephson equations refer to the phase difference $\phi = \phi_1 - \phi_2$. They describe how the current I passing through the link and the voltage V across it depend on ϕ.

$$I = I_0 \sin \phi \tag{4.19}$$

and

$$\frac{d\phi}{dt} = \frac{2eV}{\hbar} . \tag{4.20}$$

There are two types of effects, usually labelled AC and DC. In the DC effect, the phase adjusts until there is no voltage across the link (equation 4.20), leaving a DC supercurrent through the link (equation 4.19). The maximum magnitude is given by the constant I_0, which depends on the coupling strength of the link.[4] The AC effects are associated particularly with equation 4.20. This equation shows that when a DC voltage V is applied across the junction, the phase winds up continuously at an angular frequency ω given by $\hbar\omega = 2eV$. This result shows that when a Cooper pair

---

4 There is an elegant derivation of equations 4.19 and 4.20 in Feynman's *Lectures in Physics* Vol. 3 Chapter 21.

(of charge 2*e*) tunnels through the weak link, there will be an energy change of 2*eV* which will manifest itself as a photon of frequency ω.

There is much fascination in these effects, since they demonstrate the coherent nature of the superconducting wave function on a macroscopic scale. And, in addition, they open up a whole range of new devices which have revolutionised measurement science in a number of ways, some of which we now briefly introduce.

1. The RF SQUID   The mnemonic SQUID stands for Superconducting QUantum Interference Device. The RF SQUID consists in essence of a loop of a superconductor broken by a single weak link, as illustrated in Figure 4.15A. Thus there are now two conditions to take into consideration. One is that the phase around the full loop must join up, i.e. the phase difference must equal $2n\pi$ with n integral. But the other is that the phase difference across the link must be given by equation 4.19. Consider the response of the SQUID ring to an externally applied magnetic flux (Figure 4.15B). If the link is so weak as to effectively open (corresponding to $I_0=0$), then we know the answer – the flux in the ring will be precisely the applied flux, as labelled (1). If on the other hand the link is so strong that the ring is effectively solid superconductor (I_0 large), then we also know the answer, labelled (2). The flux remains constant at $n\Phi_0$ where $\Phi_0=h/2e$ is the flux quantum, since the ring is able to develop large enough supercurrents to screen out external influence.

However, for a suitable value of I_0 the equations give something like the tortuous curve labelled (3). What happens here is that as the applied flux is increased, the screening current can only increase until (at point A, for example) it equals I_0, at which point a flux quantum slips in through the weak link, allowing the screening current to relax to a manageable value (at point B). Now it is important to note that this system shows hysteresis, in that a reducing applied flux, the jump back out of the flux quantum occurs between points C and D.

Thus our SQUID can be used as a very sensitive magnetometer. A radiofrequency (hence RF) flux of variable magnitude is applied to through a resonant circuit. The amplitude at which hysteretic (and therefore lossy) behaviour takes place depends on the magnitude of the DC background flux present, so that the electronics can be organised so as to give a signal which depends strongly on this background. The signal is of course periodic in the flux quantum, but can readily resolve flux to $10^{-4}\Phi_0$ under most conditions. Thus the SQUID can be used to detect and measure DC flux or low frequency AC flux. With the use of suitable input coils ("flux transformers"), the technique can be adapted to a wide range of different measurements. This is now a standard device for magnetic and electrical measurements in many areas of science, medicine and surveying.

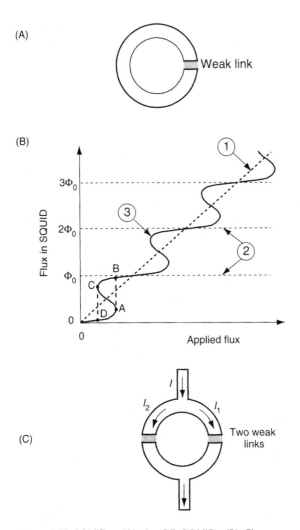

Figure 4.15 SQUIDs. (A) An RF SQUID; (B) Flux response of an RF SQUID; (C) A DC SQUID.

2. The DC SQUID With modern fabrication techniques, the favoured type of SQUID nowadays is the so-called DC SQUID. This consists again of a SQUID ring, but the ring contains two almost identical weak links (hence the need for precision in fabrication). See Figure 4.15C. A current I passing into the ring splits into I_1 through one link and I_2 through the other. When a background flux is present, and hence a circulating current in the ring, this split has to take account of the requirement of 4.19 in each link. This affects the critical current for the combined ring, which varies with applied flux as a strongly periodic function of Φ_0. There is a maximum critical current equal to $2I_0$ when the applied flux is an integral number of quanta, since $I_1 = I_2$ and the phase across each junction is the same. A minimum occurs when there is an odd half number of quanta. Mathematically, the phases of the two links combine in a way analogous to a two-slit interference pattern, giving meaning to the SQUID mnemonic. In practice the DC SQUID has a better bandwidth performance than its RF counterpart.

3. *Microwave devices and a voltage standard*   The AC effect, equation 4.20, shows that Josephson junctions can be used as detectors and mixers for microwave radiation. For instance, it is not too difficult to show that if the voltage V applied to the junction consists of a DC bias V_0 and an AC component $V_1 \cos \omega t$, then the current I will time-average to zero except when $2eV_0 = n\hbar\omega$. Thus, if one measures the DC I–V characteristic in the presence of microwave power, then steps appear in the characteristic at these voltages. Physically, one is observing photon-assisted tunnelling, namely tunnelling of a Cooper pair with the absorption of n microwave photons.

This type of effect can be used to investigate the value of the fundamental constant e/h, since the steps are such robust features. Frequency can be measured with great accuracy (say 1 in 10^8), and the voltage V_0 can be determined by comparison with (less accurate) voltage standards. Amazingly, it is found that the value of e/h is very accurately the same as determined by any superconductor and in a wide variety of geometries and weak links. The simple theory seems remarkably robust, i.e. the Cooper pair really gives a charge of integer 2 times e. Therefore, nowadays, the AC Josephson effect is used, as provider of a readily portable and universal device, to *define* the voltage standard, *assuming* a fixed value for e/h.

4.4.7 Unconventional superconductors

As a last section of this chapter, we open the lid of a very large and exciting box. The early history of superconductors entirely concerned metals, and our understanding of metallic superconductivity based on the BCS theory has formed the basis for this chapter.

But, mainly since about 1980, whole new classes of superconductors have been discovered. The most famous breakthrough resulted in the 1987 Nobel prize for Bednorz and Müller "for their important breakthrough in the discovery of superconductivity in ceramic materials". The magic substance of the 1980s was YBCO, Yttrium Barium Copper Oxide, which looked about as unpromising a conductor as you can get. It is a strongly layered material, with a complicated structure. To be superconducting it needed to be loaded with just the right amount of oxygen, enough to stop it being an insulator, but only just. At room temperature it is a very bad conductor. But on cooling merely to 94 K the black messy powder becomes a superconductor. This contrasts remarkably with the 1970s "high" temperature superconductors which were intermetallic compounds such as Nb_3Sn with T_C at around 20 K.

This breakthrough has caused interest and heartache to both engineer and physicist. Optimists (and journalists) initially thought that room temperature superconductors, magnetic levitation and loss-free power transmission etc. were just around the corner. They were wrong. Winding magnets from black ceramic powder is non-trivial; and T_C values struggle still to exceed 100 K. Furthermore, a full theory of what causes the superconductivity remains elusive. Theorists have lots of opinions, but the number of these opinions probably outnumber the number of theorists!

One fact seems secure, namely that Cooper pairs are still operative. Quantum effects were discovered very early on, since it was all too easy to make weak links in these difficult materials. And the measured flux quantum remains at the value $h/2e$, an unambiguous indicator that the superconducting "particles" carry pairs of electronic charges. (In the case of YBCO and other cuprates, the carriers are strictly holes rather than electrons.)

Another clear fact is that the materials are strongly type II in character, hardly surprising because of the strong scattering (reducing ξ) and small number of carriers (increasing λ). So flux lines abound, and show interesting behaviour since they are accessible at sufficiently high temperatures to make them mobile. Thus there are "flux liquid" states near the transition, with phase transitions to other flux arrangements.

The cuprate and allied materials, as already stated, are marginal materials in the sense that they are almost insulators, and also not far from being magnetic. The carriers (holes) are strongly correlated, so that it is unclear how far we can push the one-particle approach which has been so successful in our visualisation of metals. So far there is no agreement about the mechanism for Cooper pair formation. It seems that the standard BCS phonon mechanism will not do, since the T_C values are so high. So far the mechanisms explored include ones based on magnetic properties, including spin fluctuations; on the influence of structural distortions, static phonons perhaps by another name, leading to inhomogeneity in the complicated structures of these materials; on phases in which spin and charge of the holes order differently. All very interesting, but rather unsatisfactory.

However, there is one pointer towards a mechanism which any theorist must account for, although even this remains somewhat controversial. In standard BCS theory, we have seen that the pairing mechanism is such that electron states $+\mathbf{k}\uparrow$ and $-\mathbf{k}\downarrow$ form a Cooper pair. In terms of the pair properties, this is referred to as $S=0$, $L=0$ pairing or s-wave pairing. The

$S=0$ tag refers to the pairing of opposite spins, whereas $L=0$ refers to the lack of angular momentum of the pair; the electrons are travelling in exactly opposite directions through each other. The s-wave notation is the same as that used in atoms to label angular momentum $L=0$. Now there is an important piece of bookwork about the quantum mechanics of identical fermions. This is that the wave function for two identical particles must be antisymmetric, i.e. it must change sign, for interchange of the coordinates of the two particles. In the usual approximation of space and spin coordinates being separable, this means that one of the coordinates (space or spin) must be symmetric and the other antisymmetric. So in our s-wave pairing, we have $L=0$ which implies a symmetric wave function, thereby leaving the spin function to be necessarily antisymmetric, i.e. opposite spins and $S=0$. This is the conventional BCS picture.

In the cuprates, however, there is evidence for d-wave pairing. This means $L=2$, still a symmetric spatial function, so $S=0$ again. But the $L=2$ function leads to a much richer wave function than the simple $\Psi = C \exp i\phi$, one which mirrors the strong angular momentum. The phase structure of the wave function can be explored by Josephson junctions, and it is this sort of evidence which suggests the d-wave pairing. This is a large and ongoing topic.

Finally we may note that unconventional superconductivity is being uncovered in other materials, particularly those which show other unusual effects, such as being close to a magnetic transition or being "heavy fermion" materials in which again strong correlations between electrons are important. As a recent example, there is good evidence that in the oxide material Sr_2RuO_4, a superconductor discovered in 1994, p-wave pairing operates. This means $L=1$, an antisymmetric spatial function and hence the symmetric $S=1$ spin function. This once more gives a richness of behaviour which is being explored. And it is of particular interest in the context of this book, since p-wave spin triplet pairing is operative in the mechanism for superfluidity in ^{3}He as we shall describe in the following chapter.

4.5 SUMMARY

1   In Section 4.1 we discussed the three basic properties which characterise superconductivity. These are zero electrical resistance; exclusion of the magnetic B-field, the Meissner effect; and flux quantisation.

2   If we assume a simple macroscopic wave function, similar to that assumed in superfluid ^{4}He then, having stirred in Maxwell's equations as in Section 4.2, we can understand these basic properties.

3   Bardeen, Cooper and Schrieffer (BCS) have provided a microscopic theory of superconductivity, Section 4.3, based on a pairing mechanism of electrons in which phonons are the go-between to provide an attraction between electrons.

4   In BCS theory there is an energy gap Δ between the ground state and the excited states, and this energy gap guarantees "superfluid" properties.

5   Superconductivity provides a model second-order phase transition (Section 4.4.2).

6   Important lengths in a superconductor are the coherence length λ (describing the stiffness of the wave function, or the size of a Cooper pair) and the penetration depth ξ (describing the decay length for a B-field as the superconductor is entered).

7   Type I superconductors $(\lambda < \xi)$ are reluctant to form normal-superconducting interfaces, leading to a coarse geometry-dependent intermediate state in modest magnetic fields.

8   Type II superconductors $(\lambda > \xi)$ form as many interfaces as possible, with the magnetic flux penetrating in individual flux lines.

9   Type II materials are useful for electromagnetic applications, such as windings for magnets.

10   Josephson effects, Cooper pair tunnelling, are a direct demonstration of the macroscopic wave function.

11   They also provide SQUIDs and other devices for a wide range of precision measurements and standards.

12   Superconductivity occurs in many other systems besides conventional metals described by simple BCS theory.

13   "Unconventional" materials (Section 4.4.7) include those which are non-metals, like the ceramic high temperature superconductors, and/or those in which the pairing mechanism is more exotic than simple s-wave $(S=0, L=0)$ pairing.

FURTHER READING AND STUDY

A useful introduction is usually found as a chapter in any general text on solid state physics, for example:

C. Kittel, *Introduction to Solid State Physics*, Wiley, 1971.

H. E. Hall and J. R. Hook, *Solid State Physics*, Wiley, 1991.

There are a number of more specialised books, for example:

D. R. Tilley and J. Tilley, *Superfluidity and Superconductivity*, Institute of Physics, 1996.

P. V. E. McClintock, D. J. Meredith and J. K. Wigmore, *Matter at Low Temperatures*, Blackie, 1984 (Chapter 4).

A. C. Rose-Innes and E. H. Rhoderick, *Introduction to Superconductivity*, Pergamon, 1978.

High temperature superconductivity is a rapidly developing field, but has many excellent resources available on the internet. For example, obtain well-organised information and follow links from [http://www.superconductors.org/].

PROBLEMS

Q4.1 A precision NMR proton spectrometer uses a superconducting magnet in the persistent mode, providing 9.6 T for a 400 MHz machine. A typical specification for the magnet is that it should remain stable to 1 part in 10^7 over a period of several years. The magnet is large, with a self-inductance of 100 H. (a) Describe the sequence of operations when the magnet is first installed. (b) Calculate the maximum possible resistance of the magnet circuit to meet the specification.

Q4.2 Explain the difference between perfect conductivity and superconductivity in relation to the Meissner effect.

Q4.3 (a) Summarise briefly the arguments given in Section 4.4.3 which suggest that a length scale $\xi - \lambda = b$, say, characterises the surface energy of an N–S interface. (b) What is particular about b for a type II superconductor? (c) Consider the laminar intermediate state in a cylinder (radius r) of a type I superconductor, as illustrated in Figure 4.5B. Explain why you would expect the thickness (d, say) of the laminae to increase with b. (d) In fact, the relation is something like $d = A(br)^{1/2}$ where A (which depends somewhat on the applied field) is of order unity. Use this to estimate d in a cylinder of radius 1 mm of pure tin ($\xi = 230$ nm, $\lambda = 51$ nm).

Q4.4 Discuss the analogy between superfluid ^{4}He in rotation and a type II superconductor in a magnetic field. What is the analogue of the Meissner state in superfluid helium?

Q4.5 Use the internet and other sources to chart the history of the highest known superconducting transition temperatures. Suggest how our world might change with the advent of usable room temperature superconductors.

Q4.6 Consider a symmetrical SIS Josephson junction. Show how the Josephson equations (Section 4.4.6) can be derived from the Schrödinger equations for the two sides, when a mixing term is introduced as $i\hbar(\partial\Psi_1/\partial t) = E_1\Psi_1 + K\Psi_2$ for side 1 and similarly for side 2. Look for solutions of the form $\Psi_1 = n_1^{1/2}\exp(i\phi_1)$ etc. where n_1 is the number density of the Cooper pairs on side 1. Show that (with the notation of equation 4.14) I_0 is proportional to K and to n.

Q4.7 Core physics topics to give useful background

  (a) Maxwell's equations in electromagnetism
  (b) Para-, dia- and ferro-magnetism
  (c) Thermal properties of electrons in normal metals
  (d) Electronic transport properties in metals

Q4.8 Possible essay topics.

  (a) Cooper pairs and BCS theory
  (b) SQUIDs and their uses
  (c) The energy gap and its measurement
  (d) The electrodynamics of superconductors
  (e) Quantised magnetic flux lines
  (f) Unconventional superconductivity

Chapter 5

Liquid ³He

In this chapter we look at the properties of superfluid ³He. As indicated in the first chapter, the recent studies of this remarkable state of matter have progressed alongside improvements in refrigeration techniques. It is a good example of the interdependence of science and technology. The initial discovery of the new superfluid in the early 1970s at temperatures of about 2.5 mK used the Pomeranchuk effect, which was operating at around its low temperature limit. This gave a great motivation for the rapid development of the new technology of dilution refrigeration, coupled with nuclear demagnetisation cooling. As a result of the technical advances, many new phenomena in basic science were discovered, which led in turn to further refinements in cooling technology. And so on!

5.1 INTRODUCTION

We have already seen (Chapter 2) that liquid ⁴He becomes superfluid below 2.17 K, the λ-point. The ⁴He atom is of course a (composite) boson, having an even number of particles (2 neutrons and 2 protons in the nucleus and 2 orbital electrons). The superfluidity is related to the Bose–Einstein condensation expected in a Bose gas, arising from the symmetry of the wave function. It is a major effect which thus occurs at a "high" temperature.

On the other hand the ³He atom is a composite fermion, with only one neutron in the nucleus rather than two. Hence it has a nuclear spin of 1/2. It is in quantum mechanics written in capital letters that nuclear spin can so dramatically affect the gross properties of the substance. Even the existence of a superfluid state of liquid ³He was uncertain until it was discovered, since the only possibility is a subtle pairing mechanism, analogous to that in a BCS superconductor. A consequence of the "weak" mechanism is that the transition takes place at temperatures a thousand times lower than in liquid ⁴He.

5.1.1 A historic discovery: the A and B phases

The ground-breaking (and eventually Nobel prizewinning) discovery of superfluidity in ^{3}He by Lee, Osheroff and Richardson at Cornell University was made in a very simple way. Simple in principle, that is. The technique of Pomeranchuk cooling (see Section 3.1.2) by conversion of liquid ^{3}He to solid by application of pressure requires much ingenuity. In their experiment, the pressure was applied by a remote ^{4}He bellows system, and the melting pressure in the ^{3}He Pomeranchuk cell was monitored. This pressure measurement is essentially a temperature measurement, derived from the solid–liquid curve of the ^{3}He phase diagram. When the volume of the cell was swept uniformly as a function of time, reproducible little glitches were seen on the melting pressure response. They called these glitches the A feature and the B feature. Their curve is reproduced schematically in Figure 5.1, which indicates pressurisation (cooling) of the cell followed by depressurisation (warming).

One of the motivations for the Cornell experiment was to look for phase transitions in the solid ^{3}He in the Pomeranchuk cell, and thus it was at first thought that the A and B features were signatures from the

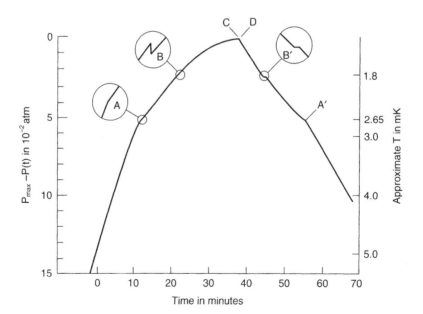

Figure 5.1 The original data of Lee, Osheroff and Richardson which indicated the discovery of the A- and B-phases of superfluid ^{3}He.

Sources: Lee, D. M. (1997) *Reviews of Modern Physics*, **69**, Extract from Nobel lectures in physics 1996, Figure 3. © The Nobel Foundation 1996.

solid. But it was rapidly realised that, no, they were from the liquid. There was a second-order phase transition at A from normal fluid to what became known as the A-phase, evident from the sudden change in slope at A. This arises in a classic second-order transition, just as with a superconductor, with an upward jump in specific heat capacity C at the transition. This jump arises in a disorder-order transition from the onset of ordering which means that the entropy S falls rapidly below T_C, hence causing the increase in $C = T(dS/dT)$. This behaviour of C for liquid ^{3}He has since been observed by many others. On the other hand, the B feature showed all the characteristics of a first-order transition, in that there was a clear latent heat involved (hence the plateau on warming), and the transition showed supercooling and thus hysteresis (seen on cooling). This suggests a phase change to a new phase, now called the B-phase.

The measurements being made at Cornell were of conventional NMR, measured by a continuous wave method. In measurements of this sort, a steady field is applied in (say) the z-direction, causing the nuclear spin levels to split (the Zeeman effect) with an energy separation of $2\mu B_z$ where μ is the z-component of the magnetic moment of the ^{3}He nucleus. A transverse RF field is applied in the x–y plane, and resonant absorption of RF power of frequency ω_L is observed when the condition $\hbar\omega_L = 2\mu B_z$ is satisfied. This frequency is called the Larmor frequency, and corresponds to a frequency of 324 kHz in a field of 10 mT. What they observed was that, as the A-phase was entered, an additional satellite line at higher frequency gradually split off from the main peak. The satellite then abruptly disappeared when the B-phase transition took place. Furthermore, this satellite line was shown to be associated with the liquid, since they were able to perform basic NMR imaging of their cell by having B_z vary in space. The satellite line was early evidence that the A-phase was strongly magnetic, so that the effective field seen by a nucleus was increased above the externally applied field. Furthermore, such an effect is absent in the B-phase. The significance of these discoveries was expounded with astonishing rapidity by Tony Leggett, then at the University of Sussex, who recognised that such properties could exist in a superfluid containing Cooper pairs with p-wave pairing.

5.1.2   Phase diagram of the superfluid

The full phase diagram for ^{3}He at millikelvin temperatures is sketched in Figure 5.2. Figure 5.2A shows the phase diagram in zero magnetic field. We now know that there are just the two superfluid phases, the A-phase and the B-phase. The Cornell A and B features occurred at the intersections with the solid–liquid line. There is considerable pressure dependence, with T_C being a factor of about 2.5 higher at high pressure than it is at

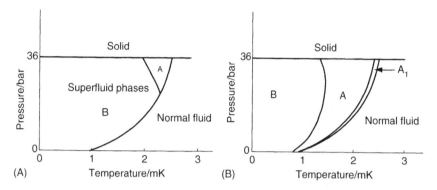

Figure 5.2 (A) The phase diagram of ³He at millikelvin temperatures, showing the superfluid phases in zero applied magnetic field. (B) Schematic phase diagram of ³He in an applied magnetic field.

zero pressure. The more dense ³He at high pressure also evidently favours the A-phase, rather than the B-phase.

On going from the normal liquid phase (N) into the superfluid, there is an upward sudden jump in heat capacity, characteristic of a second-order transition, just as in the BCS theory of superconductivity (see Section 4.4.2). The value of $\Delta C/C = (C_S - C_N)/C_N$ varies with pressure, being about 2 close to the melting pressure, falling with pressure to around the BCS value of 1.43 at 0 bar.

The transition from A-phase to B-phase is a classic first-order transition, showing evidence of nucleation problems and therefore hysteresis of the transition. There is a latent heat associated with the transition, providing warming when the A to B transition takes place and cooling when passing from B to A. The nucleation of the B-phase remains a topic of much interest, requiring the intrusion of external noise (e.g. vibration or cosmic rays) under some conditions.

When a magnetic field is applied, the phase diagram changes dramatically, as illustrated in Figure 5.2B. For low fields, the first casualty is the so-called poly-critical point, the point where the second-order normal to superfluid line is joined by the first-order A–B transition line. The curve becomes rounded there and a thin sliver of A-phase is interposed between the normal and B-phases. As the field is increased, the A-phase region grows at the expense of the B-phase. In addition a third new superfluid phase occurs at the transition, called the A_1-phase. This phase occupies a width in temperature at any given pressure which is proportional to the field. Finally, when the field is raised above about 500 mT, the B-phase disappears entirely, leaving a still thin sliver of A_1-phase with the majority of the superfluid being A-phase.

5.1.3 Magnetisation and NMR

Two other experiments are so basic to our understanding of the super-fluid that we single them out for brief mention here. One is magnetic susceptibility and the other is the NMR behaviour, alluded to above.

Since ^{3}He has only a nuclear magnetic moment, the magnetic suscepti-bility is small. Nevertheless it can be measured by static methods since it is also an ideally pure substance at low temperatures, as well as by dynamic methods using the strength of an NMR signal. Since the normal liquid can be described as a Fermi gas (see the next section) with a Fermi temperature well above the ambient temperature, it should be no surprise that the measured susceptibility χ_N of the normal phase is independent of temperature. When the liquid enters the superfluid A-phase, in spite of remarkable changes in many of its properties, the magnetic susceptibility is virtually unchanged, i.e. $\chi_A = \chi_N$. Evidently, the pairing involved in the superfluidity has no influence on the ability of the spins to flip under the influence of a small applied magnetic field. In marked contrast, when the B-phase is entered, the susceptibility drops. When the temperature is well below T_C, the susceptibility has fallen by a factor of roughly 3, i.e. $\chi_B \simeq 1/3\chi_N$. This indicates the lack of freedom for the spins in this state, so that it is not too surprising that large enough fields favour the A-phase above the B-phase as the stable superfluid state.

In the previous section, we noted how Lee, Osheroff and Richardson had observed a shifted NMR line in the A-phase using the conventional transverse NMR geometry. By measuring at different applied magnetic fields, and thus different frequencies, it was shown that the effective field seen by the spins in the A-phase is not just the applied field B_z. Instead it becomes equal to $(B_z^2 + B_{int}^2)^{1/2}$, where the magnitude of the Pythagorean addition B_{int} is independent of the applied field. Equivalently one can say that the line is shifted upwards from the Larmor frequency ω_L to $(\omega_L^2 + \Omega_A^2)^{1/2}$. This effective field increases as the temperature is lowered and the superfluid state becomes stronger. It is a large effect, with B_{int} being of order $3\,\mathrm{mT}$. The Pythagorean addition of the fields suggests that B_{int} is perpendicular to the applied B. Therefore, this also suggests the existence of a new type of NMR signal, a longitudinal effect in which the RF power is applied parallel to the external steady field. Such an effect does not exist in a normal substance, but does do so in the A-phase and it occurs at the same frequency Ω_A as above. The message here is that the small magnetic dipolar interactions of the spins all act in concert in the superfluid to give a significant force on the spin system which tries to couple it to the orbital motion of the Cooper pairs. One very smart way to observe the longitudinal resonance is simply to step the longitudinal field. This then results in "ringing" a gradually decaying signal at frequency Ω_A.

Finally, let us note that there are also many unusual effects in the NMR response of the B-phase. Although it was not observable in the first experiments of Lee, Osheroff and Richardson, ^{3}He-B also gives a longitudinal resonance, at a frequency Ω_B which is of the same order of magnitude as that in ^{3}He-A. For some orientations of the texture (see below), there are corresponding upward shifts in the transverse resonance frequency, so that NMR can be used as a tool to study B-phase textures. But in addition, at temperatures well below T_C, very long lived magnetic states in ^{3}He-B have been discovered. These can give "free induction decay" signals lasting long enough to make and drink a cup of tea! We shall return to these effects briefly in Section 5.5.

5.2 SOME THEORETICAL IDEAS

We now turn to our current understanding of superfluid ^{3}He. It is a fascinating state of matter, since it combines ideas of Cooper pairs of fermions similar to BCS, but with p-wave pairing which gives many new degrees of freedom to the system. The fact that the Cooper pairs have spin means that it is a strongly magnetic system. The fact that, as we shall see, the superfluid order parameter has vectorial character means that this anisotropic ordering can be aligned by boundaries and by applied forces, leading to textural effects analogous to those in liquid crystals. The fact that it is ideally pure means that at temperatures well below T_C the very sparse particles of the remaining normal fluid behave ballistically, giving both simplicity and interest to the dynamics of the fluid in this region.

What makes thinking about what goes on in superfluid ^{3}He at all possible to one-brained humans is that ^{3}He can be modelled with reasonable accuracy as an almost ideal Fermi gas. This arises from the amazing success of Landau's Fermi liquid theory.

5.2.1 Landau's Fermi liquid theory

Here we shall just outline the essence of Landau's results, leaving fuller treatments to various books referenced at the end of the chapter. The basic idea is that, although the interactions between atoms are very strong indeed, the interacting fermion system can be redescribed in terms of "quasiparticles" which are themselves weakly interacting. Furthermore there is (virtually) a one-to-one correspondence between atoms and quasiparticles. Hence one can use all the intuition gained from thinking about ideal gases, and apply this to quasiparticles without, usually, doing great injustice to the physics.

Landau's model uses the philosophy of the excitation picture, which deals in differences in energy between the ground and excited states. But it goes further and is built on the idea that only *changes* in energy are observable quantities, whereas the absolute value of an energy is not. Thus it works out the energy of a quasiparticle in state **k** using the difference in energy of the system when the quasiparticle distribution function is changed. The final step is to express the difference (changes again!) in quasiparticle energy between two situations, namely (1) when the one solitary quasiparticle is present (i.e. the equilibrium value at $T=0$) and (2) when other quasiparticles are present. This difference is characterised by an integral over an interaction function $F(\mathbf{k}, \mathbf{k}')$, where **k** is the state of interest and **k**′ represents other quasiparticle states.

In liquid ^{3}He we are dealing in practice with an isotropic substance which is at low temperatures so that all excitations occur virtually at the Fermi sphere. The interaction function can thus depend only on the angle θ between **k** and **k**′. Hence it can be expanded as orthogonal Legendre polynomials to give

$$F(\theta) = F_0 + F_1 \cos\theta + F_2 \left(\frac{3\cos^2\theta - 1}{2} \right) + \cdots$$

Thus all the interactions are rolled into a set of these F numbers, called Landau parameters. Or rather, there are two sets of numbers, to allow for the fact that the interactions are spin-dependent. One set is written F_0^s, F_1^s, F_2^s, etc. corresponding to the situation where **k** and **k**′ have the same spin. The other, F_0^a, F_1^a, F_2^a, etc. corresponds to **k** and **k**′ having opposite spin. All results for the non-interacting Fermi system are recovered by setting F to zero.[1] Landau's theory is found to work almost alarmingly well in normal liquid ^{3}He at low temperatures. Although the interactions are large, it is found that in practice most experimental properties can be described fairly well using only three of the Landau parameters, F_0^s, F_1^s and F_0^a. Let us mention briefly a few of these properties in the normal phase of liquid ^{3}He.

1. Specific heat capacity   This is a property which for a degenerate Fermi gas is linear in T and proportional to the density of states at the Fermi level. The heat capacity in normal liquid ^{3}He is indeed linear, but the magnitude is much larger than would be expected for a Fermi gas of the same density. This enhancement is often described as an increased effective mass of the quasiparticles above that of a free ^{3}He atom. For

---

1 Beware that you will find different notations for Landau parameters in the literature, particularly for the antisymmetric spin-changing series. These numbers are variously written as Z, G, $4F''$ or (regrettably) F''. I confess to having deliberately omitted the rogue factor 4 in the definition of F_0^a in my *Statistical Physics*. The more usual definition ($F_0^a = Z_0/4$) is used in this chapter.

example at zero pressure, the heat capacity is found to be a factor of about 2.8 higher than the ideal gas value, meaning that $m^*/m_3 = 2.8$. Landau theory gives a ratio of $m^*/m_3 = (1 + F_1^s/3)$, suggesting therefore that $F_1^s = 5.4$. At higher pressures, the effect is even larger, with an enhancement of over 5.8 and a value of F_1^s greater than 14. Remember that a non-interacting assembly would have all $F = 0$, so we see that ^{3}He is a very strongly interacting system.

2. Magnetic susceptibility   The specific heat enhancement arises from an enhanced density of states at the Fermi level. Thus it is a factor also in the Landau theory for the magnetic susceptibility. However, the magnetisation depends on turning over of spins of the quasiparticles, so an additional factor of $(1 + F_0^a)^{-1}$ appears. This factor can therefore be determined experimentally from the ratio of susceptibility to heat capacity. It is found to equal about 3.3 at zero pressure and not to depend much on pressure. This implies a value of F_0^a equal to about −0.70 (rising slightly to around −0.77 at the melting curve). We should note that this implies a very strong magnetic interaction, tending towards nuclear ferromagnetism. Indeed a value of $F_0^a = -1$ would give infinite susceptibility, i.e. spontaneous magnetisation, and such a liquid would be ferromagnetic. So we learn that ^{3}He is three quarters of the way there towards ferromagnetism. This strong interaction is important as we shall see below in the formation of Cooper pairs.

3. Sound modes   The velocity of sound c_1 is found theoretically to be given by $c_1^2 = (1 + F_0^s)p_F^2/(3m_3 m^*)$, where p_F is the Fermi momentum. Hence c_1 depends again on the value of F_1^s, but it also has the factor $(1 + F_0^s)^{1/2}$. Not surprisingly, the interactions make the interacting fluid very much stiffer (i.e. less compressible) than the corresponding gas. This is a very large effect with F_0^s being about 9 at zero pressure, rising sharply with pressure as the atomic volume falls, to be around 90 at the melting curve.

In addition to this normal first sound, Landau predicted a further effect arising from these interactions. In spite of the high density of the fluid, mean free paths in liquid ^{3}He are long, and become longer as temperature falls. This arises from the exclusion principle. The only mechanism for quasiparticle scattering in the bulk liquid is for one quasiparticle to scatter from another. For two quasiparticles to scatter, there must be empty states of suitable energy and momentum for both to enter. This restricts scattering to quasiparticles and states within $k_B T$ of the Fermi level, and so introduces two factors of $k_B T/E_F$ into the scattering rate. Thus the mean free path ℓ varies as T^{-2}. At low temperatures and high enough frequency, it is therefore quite possible to reach a collisionless regime in which ℓ is greater than the wavelength of the sound wave. In a non-interacting system, there is no propagating sound mode under these conditions; sound does not pass through a very low pressure gas. However, in the interacting system, a new sound mode becomes possible, in which the restoring stiffness of the degenerate Fermi sea comes entirely from interactions. Landau named the new mode "zero sound", and its discovery is one of the jewels of Landau's theory. It has usually been observed as a temperature-dependence of the sound propagation at a fixed ultrasonic frequency, say 50 MHz. At high temperatures, a strong first sound mode (velocity about 180 m s^{-1}) is observed, whereas at low temperatures we have the new zero sound (at about 190 m s^{-1}). Even more strikingly, there is very strong attenuation of the wave around the change-over temperature, a characteristic of a system which cannot decide between two possibilities.

4. Other properties   These three parameters, without further modification, give a description of other properties also. The transport processes of viscosity, diffusion, spin diffusion, thermal conduction, can all be explained quantitatively without any further major assumptions. If you are very particular, the higher Landau parameters must also be included, but these corrections in the normal fluid are comparatively minor. [They can become somewhat more significant in the superfluid since additional anisotropy is introduced by the p-wave pairing.]

To sum up this section, we have seen that the properties of the normal liquid ^{3}He can be understood in terms of the almost ideal Fermi gas model, so long as the temptation to use the numerical parameters of the free gas model is avoided. The interactions are in fact very strong, and the independent entities, the quasiparticles, have dynamical properties very different from those of the ^{3}He bare atoms. This surprisingly simple result arises in the degenerate limit of Fermi statistics only, where the major action is restricted to be close to the Fermi surface. In many ways it is analogous to the success of the one-electron picture of metals. It is certainly worth reminding oneself that in liquid ^{4}He there is no such simple crutch. In that case, the interactions dramatically modify the qualitative behaviour of the liquid when compared to the Bose–Einstein condensation of an ideal gas.

5.2.2   p-wave pairing

Having seen that simple intuition based on ideal gases can work for the normal fluid phase of ^{3}He, we are now equipped to talk about the nature of the superfluid.

As we have noted, it is now widely accepted that p-wave pairing occurs. This is strongly suggested since:

1   An ideal looking second-order transition visible in the heat capacity of the Fermi system suggests formation of "weak" Cooper pairs by analogy with conventional BCS superconductivity.

2   Experimentally there are magnetic properties, ruling out $S=0$, since the simple BCS s-wave pairing ties up opposite spins in a non-magnetic pair.

3   For ³He, it might somewhat stretch one's credulity to imagine pairs having $L=0$, even for a true believer in the indistinguishability of identical particles according to the ideas of quantum mechanics. This is because $L=0$ means no angular momentum of the pair, so that they must pass right through each other (electrons, point particles, yes; ³He, enormous atoms, no !?).

4   In any case, if we accept that $S=1$ from the observed magnetic properties, then we must have L odd; and $L=1$ is by far the simplest option. [This result comes from the wave function symmetry for identical fermions; the total wave function must be antisymmetric for co-ordinate exchange, $S=1$ means a spin symmetric function, so the spatial wave function must be antisymmetric, and that means L odd.]

When we decide on p-wave pairing, the theoretical situation is fascinating, yet very complex and quite a challenge.

In s-wave theory, as in our earlier discussion about superconductivity (Chapter 4.3), there is just one sort of Cooper pair, constructed from the $\mathbf{k}\uparrow$ and $-\mathbf{k}\downarrow$ one-particle states. The pair is a spin singlet and an orbital (spatial) singlet. The superfluid ordering can be described by an order parameter, the macroscopic wave function of the form $\Psi = C \exp(i\phi)$ with C and ϕ simple scalars. This is the recipe we have adopted in discussing both superfluid ⁴He and (conventional BCS) superconductivity. We have seen that it leads to a simple isotropic energy gap Δ between the super-fluid ground state and the excitations.

But now, in p-wave pairing, we have triplets in both space and spin coordinates. There are three possible spin configurations, giving $S_z=+1$, $0, -1$. In an obvious terminology, these spin states may be referred to as $\uparrow\uparrow$, $\uparrow\downarrow$ and $\downarrow\downarrow$. The $S_z=0$ state, here written simply as $\uparrow\downarrow$ is of course the symmetric spin pairing more fully written as $1/\sqrt{2}(|\uparrow\downarrow\rangle + |\downarrow\uparrow\rangle)$. We also have three orbital configurations corresponding to $L_z=+1$, $0, -1$. There-fore there are in all no fewer than 9 types of Cooper pair, and the order parameter must in general be written as a 3×3 matrix, $A_{\mu j}$ where μ covers the three spin states and j the three orbital states. Since each of the As is a complex number, that means that a Cooper pair has 18 numbers associated with it.

In principle, therefore, there is a fantastic variety of possibilities for the most stable state. Fortunately, however, there is a simplification in that the various spin components in the Cooper pair wave function behave almost independently. Thus the observed magnetic properties of the A, B and A_1 phases can be used to give important clues about the make-up of these phases.

1   We saw that the A-phase susceptibility is the same as in the unpaired normal state. This is a signal that only $\uparrow\uparrow$ and $\downarrow\downarrow$ states are involved, since the $\uparrow\downarrow$ state has no magnetic properties. Thus the A-phase is referred to as an "equal spin pairing" state.

2   The A_1-phase only occurs near T_C and when a magnetic field is applied. It is thought to involve $\uparrow\uparrow$ spins only, i.e. those spins aligned to the applied field.

3   The B-phase susceptibility is lower than that in the normal state, a signal that the $\uparrow\downarrow$ state is now involved. In fact it is believed that all three spin components are equally involved. At first sight this might seem to imply that the susceptibility should reduce by a factor of 2/3 rather than the observed factor of about 1/3. However the interactions, as described by the Landau parameters, need to be considered. In the normal fluid, we saw above that χ_N is enhanced by the large ferromagnetic interaction measured by F_0^a. But this spin-dependent interaction is decreased when non-magnetic $\uparrow\downarrow$ pairs are involved, and hence the theory based on Landau's Fermi liquid theory turns out to give good agreement with the observed $\chi_B \approx 1/3\chi_N$.

5.2.3   Nature of the A- and B-phases

We are now in a position to describe the nature of the superfluid ordering in the A- and B-phases.

But first there is an amusing subtext. The discovery of superfluidity in ^{3}He is one of the most exciting advances in physics in the last third of the twentieth century; it is truly a new state of matter. So, it is easy to be somewhat scathing about the imagination of low temperature physicists, who called the phases A and B rather than "top-charm" and "bottom-charm" or something of the sort. Of course, one excuse is that the A and B features were named and discussed before it became clear what their origins were. Another is that it was a flash of prophetic insight, designed to honour the dedication of theorists. This is because, independently and 10 years before superfluid ^{3}He was discovered, theorists had been speculating on the possible states of a p-wave superconductor. Two such states were described in some detail (1) by Anderson, Brinkman and Morel[2], or ABM, and (2) by Balian and Werthamer, or BW. And it just so happens that the A-phase is the ABM state; and the B-phase is the BW state. Very neat. So perhaps Lee, Osheroff and Richardson were minding their Ps and Qs after all!

---

2 To be accurate, the 1960s paper was by Anderson and Morel, and the ideas were expanded further in 1973 by Anderson and Brinkman.

The A-phase

First, let us consider the A-phase. As we have seen, only ↑↑ and ↓↓ states are involved in this superfluid. Therefore three of the nine elements of $A_{\mu\nu}$ vanish, and it is not surprising that a simpler description is possible. In fact the ABM state shows very high anisotropy. The wave function can be taken as the product of an orbital part and a spin part. In this phase, all the Cooper pairs share the same direction of their orbital angular momentum, conventionally written as a unit vector **l**. The spin part is characterised by another unit vector **d**, which again is the same for all Cooper pairs in the A-phase. The vector **d** is defined as the direction of *zero* spin projection, so that it is perpendicular to the direction which defines the up and down of the ↑↑ and ↓↓ spin wave functions.

Now the ABM state, corresponding to the equilibrium state without any external constraints, has these two vectors **l** and **d** very firmly parallel to each other. With hindsight (or with Leggett's intuition) it is not too difficult to see why. Consider two parallel nuclear spins rotating about each other, a fixed distance apart. Parallel magnetic moments will tend to repel each other, so there is a small energy penalty in the alignment. [I say small because the nuclear moment is itself small, which is why nuclear spin systems tend to remain as ideal paramagnets typically to nK temperatures.] The two extreme cases for the geometry of the rotating pair of spins are (1) where the spins are aligned parallel to the axis of rotation, and (2) where the spins are aligned perpendicular to the axis of rotation. We can use a simple classical argument to suggest that case (2) will minimise the effect of the repulsion. In case (1), the magnetic moments are always perpendicular to their line of separation, the configuration which maximises the repulsive energy of the two parallel magnets. On the other hand, case (2) allows the magnetic moments sometimes to be parallel to the line of separation, a more favourable configuration. In (helpful) jargon, this is referred to as "spontaneously broken spin-orbit symmetry". What it means is that there arises a coupling between the vector **l** describing the orbital motion and the vector **d** describing the spins. The vector **l** is the axis of orbital rotation, whereas **d** represents a direction of zero spin projection so that the spin direction must be perpendicular to **d**. Hence we see that the condition **l** ∥ **d** satisfies case (2).

An aside comment on order-disorder transitions may be helpful here. What happens is that when it is cooled through T_C the system in question is forced to make a choice, i.e. it is forced to break a symmetry. The transition to ferromagnetism is the easiest to visualise. In the paramagnetic state in zero magnetic field there is no preferred spin direction. But, when the ferromagnetic state is formed, a particular spin alignment is singled out and a spontaneous magnetisation appears. A symmetry (spin rotation) is spontaneously broken. Similarly in antiferromagnetism. In an s-wave pairing superfluid, the order is characterised by its wave function with a specific phase ϕ, coherent over a long range, a feature which did not exist above T_C. Along with this is the growth of an order parameter Δ. In jargon, this involves breaking "gauge" symmetry, associated with spatial properties. In superfluid ^{3}He, both of these types of symmetry are broken together. As we have seen both spin and spatial ordering occur in the superfluid state. But what the previous paragraph means is that they are not separate independent effects, but there is also a correlation between them, a further symmetry breaking constraint on the system, the spontaneously broken spin-orbit symmetry.

To return to A-phase, the order parameter of the ABM state corresponding to the space and spin alignments introduced above can be written as

$$A_{\mu j} = \Delta\, d_{\mu}(m_j + in_j) \tag{5.1}$$

where the orbital pairing is included in terms of two perpendicular unit vectors $\mathbf{m}$ and $\mathbf{n}$ which form an orthogonal triad with the l-vector. This is a correct but not very transparent expression, perhaps. Note that although it contains no explicit phase factor, rotation of $\mathbf{m}$ and $\mathbf{n}$ about l by an angle ϕ is the same as multiplication by $\exp(-i\phi)$.

The important feature to grasp here is that we are describing a very strong anisotropy of the superfluid ordering in the k-space of the ^{3}He quasiparticles. The ABM state has its strong axis, the direction of both orbital (l) and spin (d) ordering. [other name for the state, naturally beginning with A, is the axial state.] The order parameter is usually described as an anisotropic energy gap, and is given by

$$\Delta(\mathbf{k}) = \Delta_0 \sin\theta \tag{5.2}$$

In equation 5.2, the angle θ is the angle between the symmetry axis of the order and the direction $\mathbf{k}$ of a quasiparticle. The gap is illustrated in Figure 5.3. It is perhaps not too hard to see where this gap anisotropy comes from. A quasiparticle state with its momentum $\mathbf{k}$ aligned along the axis cannot possibly combine with a $-\mathbf{k}$ partner to form a Cooper pair with an angular momentum contribution along this axis. Hence these quasiparticles do not contribute to the A-phase ordering, i.e they have $\Delta = 0$. On the other hand, pairs around the equator naturally have their full angular momentum in the axial direction, so provide the maximum contribution to the superfluid state, i.e. they have maximum order parameter $\Delta = \Delta_0$. We may note that the zero gap along the axis can dominate the thermal properties of the superfluid, giving power law rather than exponential behaviour.

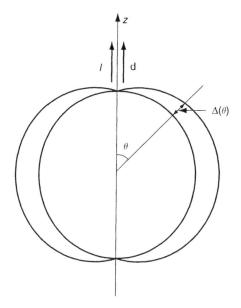

Figure 5.3 The A-phase energy gap. The diagram shows the angle dependence of the gap around the Fermi surface. The figure has rotational symmetry about the z-axis (l-direction).

The B-phase

The work of Balian and Werthamer showed that the stable state of a weakly coupled p-wave superfluid in zero magnetic field should be quite different from that indicated by the highly anisotropic ABM state, described above.

The picture of the original BW state is one which has as much symmetry as possible. As mentioned above, the Cooper pairs are of an equal admixture of all three spin states, ↑↑, ↓↓ and ↑↓. It also has an equal admixture of all three orbital states, $L_z = -1$, 0, +1. The simplest combination is such that the total angular momentum $J = L + S$ of each pair is zero. As above we may characterise a Cooper pair made up of $(\mathbf{k}, -\mathbf{k})$ one-particle states by vectors which give the direction l of the orbital motion and the direction **d** of zero spin projection. The difference now is that these directions are different for each pair, whereas in the A-phase they are a universal property. In this case l is perpendicular to **k**, since the particles are circling around each other to form the pair. In order to give $J = 0$, the spin must also be perpendicular to **k**, and hence the direction of zero spin **d** is parallel to **k**. The overall picture of $\mathbf{d}(\mathbf{k})$ for this original BW state is therefore of

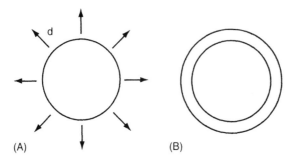

Figure 5.4 Structure of the BW state (before the rotation of 104° about a vector n is applied to describe the B-phase). (A) The d-vector around the Fermi surface; (B) the isotropic energy gap.

a spherically symmetric pincushion (hedgehog?), as illustrated in Figure 5.4. Unsurprisingly, this leads to a spherically symmetric phase, with an isotropic order parameter, i.e. the energy gap Δ is constant for all values of **k**. In the notation above, the order parameter matrix is given by

$$A_{\mu j} = \Delta \exp(i\varphi) \, \delta_{\mu j} \tag{5.3}$$

where $\delta_{\mu j}$ is the Kronecker delta.

However, there is another feature to be stirred into the argument, before we have a full picture of the B-phase. In the absence of spin-orbit interactions, this original BW state turns out to be degenerate with any state in which the spin co-ordinates are rotated with respect to the space co-ordinates in any arbitrary fashion. Hence, any state given by

$$A_{\mu j} = \Delta \exp(i\varphi) \, R_{\mu j} \tag{5.4}$$

will do equally well, where $R_{\mu j}$ is any rotation matrix. A rotation can conveniently be characterised by an axis of rotation **n** and an angle ϑ of rotation. The original BW state corresponds to $\vartheta = 0$, so that the **n** axis is immaterial. But the states with $\vartheta \neq 0$ now do show wave function anisotropy, associated with the axis **n**. We may note that a rotated state is no longer a pure $J=0$ state, but contains admixtures of $J=0$, 1 and 2. However, the rotation has no effect on the energy gap, which remains isotropic.

Finally we can inject the symmetry-breaking effect of spin-orbit interactions, which is found to favour a specific angle of spin rotation. Leggett (again!) showed that the energy change of the state depends on ϑ, being proportional to

$$2\cos^2\vartheta + \cos\vartheta + \frac{3}{4} \tag{5.5}$$

so that the minimum interaction energy is achieved when $\vartheta = \cos^{-1}$ $(-1/4) = 104°$. This state with $\vartheta = 104°$ is now believed to model the B-phase correctly. It is sometimes referred to as a pseudo-isotropic phase, in that the energy gap is isotropic, so that many properties of the excitations are dominated by simple gap Boltzmann behaviour through the factor $\exp(-\Delta/k_B T)$. Nevertheless, the underlying structure of the superfluid has anisotropy governed by **n**, and this has some remarkable effects, particularly on the NMR properties of the B-phase.

Influence of a magnetic field

Finally, we consider briefly the effects of a magnetic field. They are significant because the Cooper pairs have spin and thus a magnetic moment.

First, as mentioned above (see Figure 5.2B) there appears a new phase, the A_1 phase, in a narrow sliver at T_C. This involves the spin pairs ↑↑ only, which are aligned with the field. Hence it is a (ferro-)magnetic superfluid. The sliver depends linearly on field as does the Zeeman energy difference between ↑↑ and ↓↓. This splitting means that there are more up spins than down spins, and hence a higher density of states at the Fermi level for up spins. We can think of the ↑↑ and ↓↓ superfluids therefore as having slightly different values of T_C. The non-magnetic properties of the phase are similar to the A-phase, since it is still a state with universal axial anisotropy.

Secondly, there is the A-phase. Again, the properties are not dramatically altered, although as magnetisation takes place there is a (small) imbalance in numbers between up and down pairs. The large effect is that the A-phase starts to take over the phase diagram (Figure 5.2B) at the expense of the B-phase, until above around 0.4 T the B-phase is swallowed up entirely. This effect arises from the growing dislike of the B-phase for the field. The A-phase, having the magnetic properties of equal spin pairing, is not hindered by the magnetic field, quite unlike the response of an s-wave superconductor.

Thirdly then, there is the B-phase. Here, the magnetic field does indeed have a harmful effect on the ↑↓ pairing, in the same way as it does in the s-wave case. The point is that an $S_z = 0$ Cooper pair needs to find one-particle states around E_F for both up and down spins. However, as the Zeeman splitting comes into play, the Fermi surfaces for up and down spins separate in **k**-space. Hence the $S_z = 0$ pairing becomes weaker. On the other hand, there is no such problem with the equally paired spins, since they lie on the same Fermi surface. Hence the increasing stability of the A-phase at B's expense. [There is an interesting, but related, aside here about why the A-phase exists at all in zero field. It is thought to arise from spin fluctuations which stabilise the more magnetic state; so again it arises more at high pressures where the F_0^a Landau parameter is bigger, as the ^{3}He atoms get squashed closer together.]

Actually, there are significant distortions to the B-phase, even when it is still in the stable phase, when the magnetic field is present. The point is that, although the B-phase consists of equal numbers of the three spin states, these spin components are not evenly spread across the Fermi surface. Let us take the field direction to define the z-axis. It can be shown that the rotation axis **n** has a preferred direction parallel to the field, so that in bulk **n** also lies along the z-axis. A bit of geometry shows that the pairing states on the Fermi surface formed from states with momentum parallel to **n** must be populated by the $S_z = 0$ spin states. This is because the **d**-vector of the original BW state is directed along **k** (the pincushion effect). Here it is parallel to **n**. The 104° rotation of **d** about its own axis has no effect, so that **d** remains normal to the Fermi surface at this point. Since **d** gives the direction of zero spin projection we can see that this allows only $S_z = 0$ spin components. Hence states near the **n** direction, i.e. parallel to the magnetic field, favour ↑↓ components as stated. Similarly we can see that the equal spin states, $S_z = \pm 1$, are predominant in the $k_x - k_y$ plane, normal to **n**, since the 104° rotation about **n** leaves **d** still in the equatorial plane. Thus as the magnetic field is increased, the superfluidity is weakened near the axis and maintained at the equator, giving a gradually increasing anisotropy to the energy gap. Hence the B-phase distorts towards an axial state, and finally accepts the inevitable and becomes the even more strongly axial A-phase. Furthermore, it is worth remarking that even in distorted B-phase, the pairs which make up the superfluid develop a non-zero average angular momentum, parallel to the field direction. This can be thought of as defining an emergent l-vector parallel to the field and to **n**. This again may be thought of as a precursor of the transition to the A-phase.

5.2.4 Textures

The A-phase

We have seen that there are vector order parameters associated with ordering in superfluid ^{3}He. In the A-phase, the effect is extremely strong, with the vector **l** giving the direction of the orbital angular momentum of all pairs, and the vector **d** giving the direction of zero spin of all the pairs. We have already seen that in the bulk equilibrium state, these two directions are aligned to give a single well-defined axis. The spatial ordering described by the l-vector is similar to ordering in a nematic liquid crystal, and ^{3}He-A has many of the interesting topological and directional properties which are analogous to (say) how the figures on an LCD display work.

The first property is that there is a very strong boundary condition at a wall – and all real experiments tend to have walls. This concerns the orbital motion. Right at a wall, it is obvious that a circulating p-wave Cooper pair must have the vector 1 perpendicular to the wall, since the only reasonable plane of rotation is in the plane of the wall. A pair rotating in any other plane cannot approach it any closer than roughly the radius of rotation. This radius is of the order of the coherence length ξ of the wave function. We saw in the discussion of quantised vortex lines (Section 2.3.3) that in superfluid ^{4}He $\xi \sim 0.1$ nm, a very small length. This arises from a competition between the (very large) condensation energy and the energy penalty of bending the wave function. Here in superfluid ^{3}He the condensation energy is much smaller, so that the wave function can change only much more gently, and we find that ξ is much larger. Well below T_C in ^{3}He, ξ varies from about 80 nm at zero pressure down to about 15 nm at the melting curve.

This boundary condition by itself shows why careful experimental design is needed if a reproducible texture is required. The simple case is between two narrowly spaced parallel plates. Ignoring the edges, the texture is such that 1 is everywhere along the normal. On the other hand, for a simple spherical box there is an obvious problem. The boundary condition requires the texture (the direction of 1) to be radial at the wall. But this leads to topological problems, as illustrated in Figure 5.5. It cannot be radial everywhere. An all-radial texture would have a mess at the centre of the sphere, where there would have to be a point singularity (Figure 5.5A). Note that any such "point" singularity would involve not just a mathematical point, but a volume of dimension roughly the coherence length of the wave function. There is also the possibility of textures with a line singularity at the wall, giving a boundary between regions of 1 into the wall and 1 out of the wall, giving the texture of Figure 5.5B,

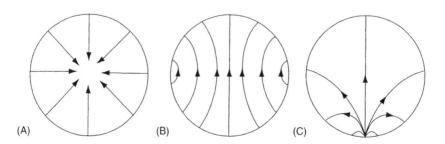

Figure 5.5 Possible l-textures in a spherical box containing superfluid ^{3}He-A. All diagrams have rotational symmetry about the vertical axis. (A) Point singularity at centre; (B) Line singularity around the equator ("PanAm" texture); (C) Point singularity at south pole ("Boojum" texture).

usually refered to as the "PanAm" texture because of its similarity to the logo of the airline Pan American. And if this line at the wall shrinks to a point, then we get another more stable texture, imaginatively called a "boojum" (Figure 5.5C), after the elusive and mystical figure in Lewis Carroll's epic poem, "The Hunting of the Snark". (Some fancy names at last!).

The above discussion relates to an ideal system in thermal equilibrium with no perturbing effects. And, as we see below, there are many perturbing effects, so that unless your experiment can be very carefully designed to give a definite stable texture, your readings can be strongly history-dependent (I speak from experience here!). Below are listed some of these perturbing effects and the influence they have on the A-phase. A lot of sophisticated calculation of energies is needed to work out these influences in detail. A measure of the strength of a perturbation is to quantify the magnitude of disturbance required for the relevant energy to match the strength of the spin-orbit coupling energy, which couples l to d, as discussed in Section 5.2.3. In any practical arrangement, the calculation must take account of another effect, in that there are energy penalties associated with the spatial distortion of the wave function caused by disturbing the equilibrium texture. This idea introduces a "bending length" which describes the minimum radius of curvature of the texture which can be expected in the presence of the perturbation. Naturally, this bending length is shorter the stronger the perturbation. In ^{3}He-A this length is typically of order $10\,\mu$m.

1. Mass and heat flow   As with ^{4}He in the two fluid region, a superfluid mass flow $\mathbf{v_S}$ can be created in a number of ways. It can be produced by mechanical means, such as by diaphragm movement. Or it can be produced by a heat flow, which results in a counterflow of normal and superfluid as discussed in Chapter 2. [The difference is that the normal fluid is highly viscous in ^{3}He, so that there is no propagating second sound mode to match that in ^{4}He.] In this case, the preferred state is with l parallel to $\mathbf{v_S}$. A flow of several mm s^{-1} would be required to match the dipole energy, a value bigger than a typical critical flow velocity.

2. Magnetic field   This is a much bigger effect, the scale of magnetic field to match the spin-orbit alignment being a few mT only. Naturally, the B-field addresses the spin system, so as to align spins parallel to the field. Hence the preferred alignment is **d** perpendicular to **B**. The fact that this perturbation can exceed the spin-orbit interaction leads to some interesting experiments. One of these (performed by the group of Hall and Hook in Manchester) uses the slab geometry mentioned above, in which l is securely locked to the slab normal. When a B-field is applied also along this normal, i.e. trying to drive **d** away from its (spin-orbit) alignment with l. Small fields have little effect, but when the field is gradually increased, there is a sudden transition when **d** decides to switch and to obey the B-field rather than the locked l-texture (a transition analogous to the sudden bending of a longitudinally compressed rod, such as a foot ruler).

3. *Electric field*   An electric field **E** tends to drive l to a perpendicular orientation. The scale field strength is a few kV per mm, the sort of value normally encountered only in ionic mobility experiments.

The B-phase

First, let us consider again the influence of a wall, when B-phase is the stable bulk phase. This is actually rather an interesting topic, since it is a good question to ask whether you can have ^{3}He-B near a wall. The hand-waving arguments given above about the strong boundary condition for ^{3}He-A seem equally valid here. Certainly a rough wall has a strong influence on the orbital angular momentum of any pair near it. Hence such a wall forces the nearby liquid to mimic aligned A-phase, which then gradually reverts to B-phase over a coherence length or so.

The effects of textures in the B-phase are usually much less dramatic than in ^{3}He-A. One reason is the weaker influence of the various applied constraints. For example, the bending length in ^{3}He-B in a typical NMR field of 30 mT is of order 1 mm (and proportional to $1/B$), 100 times larger than the corresponding length in the A-phase. In addition, textures in ^{3}He-B are in any case unimportant in many measurements, since the energy gap and hence the normal fluid density are both isotropic.

Nevertheless, there is weak vector ordering associated with **n**, the rotation axis. The axis comes about from minimising spin-orbit energy, and it is therefore not surprising that the magnetic properties such as NMR are influenced by the texture. When the calculation minimising the energy in the presence of a B-field is carried out, it turns out that the 104° rotation about **n** is still favoured and this is a strong effect (with a characteristic length of order 10 μm, similar to the A-phase). There are then fairly weak constraints (characteristic lengths of order 1 mm) which determine the alignment of the **n** vector. These are:

1   **n** prefers to be aligned to **B**, as already mentioned above.
2   The influence of a wall is to reduce the contribution of the zero spin component parallel to the surface, and it can be shown that this favours **n** being normal to the wall when the applied field is small.
3   When both field (of order 10 mT or more) and wall are present, the situation becomes more complicated. Hence, for example, working out the expected NMR lineshape (Section 5.1) is not straightforward.

There is competition between various alignment effects, and also the energy penalties associated with bending the **n** texture have to be included.

4   If there is counterflow present in addition, then we must also stir in the flow energy, which depends on the relative directions and magnitudes of field and counterflow velocity, $v_S - v_N$. All this is a formidable task, even for a professional theoretical physicist.

5.3 EXPERIMENTAL PROPERTIES OF SUPERFLUID ^{3}He

We mention briefly several experiments which, using our experience with superfluid ^{4}He and/or superconductivity, give support to the picture of the superfluid state of ^{3}He-A and ^{3}He-B outlined in the previous sections. As we shall see, there is a wide variety of possible experiments in this remarkable substance.

5.3.1 Specific heat capacity

This basic thermodynamic property gave the first evidence for the existence of the superfluid states, as discussed in Section 5.1. Later measurements of heat capacity have shown a sharp, clean second-order transition from the normal state to the superfluid state, analogous to that in superconductivity. As T is reduced through T_C there is a sudden upward jump. The magnitude of upward step is quite strongly dependent on pressure, and may be characterised by the number $(C_S - C_N)/C_N \equiv \Delta C/C$ as in a superconductor. The BCS weak coupling value of $\Delta C/C$ is 1.43. The experimental value for $\Delta C/C$ is around the BCS value at low pressure, but above 10 bar it gradually rises, until it is about 2.0 at the melting pressure (34.3 bar). We may note that this sort of variation is not unusual for superconductivity in various metals, the higher values of $\Delta C/C$ being taken as evidence for strong coupling effects.

The jump also gives support to the picture of the A_1-phase given above, the splitting of the A-phase transition in the presence of a strong magnetic field. Precision measurements show two jumps in the measured heat capacity. The higher jump has $\Delta C/C$ about 0.8, with the deficit to the full zero field value being made up by the second. This ties in with the formation of the A_1 ↑↑ phase at the upper T_C followed at a lower T_C by the formation of the usual A-phase.

The temperature dependence of C below T_C also agrees nicely with the gap structure. In the B-phase, the dependence becomes dominated by an exponential of the usual form $\exp(-\Delta/k_B T)$, where Δ is the isotropic B-phase gap. It is worth remarking that the value of the gap is about

$1.76\,k_B T_C$, again the BCS value, at low pressures, but rises to around $2k_B T_C$ at higher pressures. On the other hand, in the A-phase, the anisotropic gap with its nodes along the 1-direction means that low-energy excited states exist at values of **k** close to the nodes. At low temperatures, the number of states excited is proportional to T^2 which should lead to power-law behaviour, $C \propto T^3$. This prediction is borne out by experiment.

5.3.2 Nuclear magnetic resonance

The unusual and informative properties of NMR have already been described briefly in Section 5.1. These include the frequency shifts in the superfluid states, the existence of a new field-independent longitudinal resonance and the ability to discriminate between the A-, B- and normal-phases. We must add that there are also textural effects, since an NMR excitation involves disturbing the equilibrium state by applying a force to the spin system, hence disturbing **d**, and observing the relaxation processes back to equilibrium. Overall, NMR, studied by pulse or by continuous wave methods, is a valuable diagnostic tool for the superfluid order. We shall return to the use of NMR in the study of vortices in Section 5.3.6, and to new effects arising from the influence of superfluidity on the response of the ^{3}He spins in Section 5.3.7.

5.3.3 Mechanical measurements

In superfluid ^{4}He we saw (Chapter 2) that the mechanical flow properties played an important part in our understanding. In particular, the experiment of Andronikashvili clearly demonstrated the physical reality of the two-fluid model in terms of the normal and superfluid densities. The same sort of measurements can be made in superfluid ^{3}He. Careful measurements using a torsional oscillator can be used to study both the normal fluid density (mainly from the effective mass of an oscillator which governs its frequency) and the viscosity of the normal fluid (mainly from the damping of the oscillator). Other mechanical measurements of relevance make use of vibrating wires, which we shall discuss in some more detail below. The principal results are:

1   In the B-phase there is a well-defined scalar normal fluid density, in the same way as there is in ^{4}He.
2   In ^{3}He-A one is forced to define a vector normal fluid density, since the value $(\rho_{N\parallel})$ measured by motion parallel to the 1 vector is different from that measured normal to it, at the same temperature and pressure. In fact $\rho_{N\parallel}$ is the larger of the two, an idea consistent with the anisotropy of the superfluid energy gap which has nodes in the parallel direction.

3   In the B-phase, we have seen that the macroscopic wave function has a well-defined phase. Therefore the superfluid velocity field satisfies the same constraint, curl $\mathbf{v}_S = 0$, as in ^{4}He.

4   In the A-phase, there is no such overall phase factor, and it turns out that curl $\mathbf{v}_S$ can be non-zero if the l texture is bent.

5   At low temperatures in the B-phase, the normal fluid density falls exponentially dominated by the usual $\exp(-\Delta/k_B T)$ factor. This is well illustrated by the drop by many orders of magnitude in damping of a vibrating wire.

6   The normal fluid viscosity falls quite rapidly below T_C to about 0.2 times the normal state value, a result that is not immediately intuitively obvious!

As already implied by the success of a two-fluid approach, mass super-flow occurs in superfluid ^{3}He-B showing that it is indeed appropriate to call it a superfluid. This is to be expected, as we discussed in Chapter 1, from the nature of the dispersion relation of the excitations. The import-ant feature is that there is an energy gap Δ at the Fermi momentum.[3] For the isotropic ^{3}He-B the dispersion relation takes the same form as in a BCS superconductor, as discussed in Section 1.5.2 (see Figure 1.7) and in Section 4.3.2 (Figure 4.8). It has the form (compare equation 4.13)

$$E_p + \sqrt{(\varepsilon_p^2 + \Delta^2)} \tag{5.6}$$

where ε_p is the energy in the excitation picture of a quasiparticle in normal liquid ^{3}He with momentum $\mathbf{p}$. This suggests a Landau critical velocity equal to $v_L = \Delta/p_F$ where p_F is the Fermi momentum. This Landau velocity is quite small, about 30 mm s^{-1} at zero pressure.

We have noted that in superfluid ^{4}He simple flow (or counterflow) experiments show dissipation at velocities very much less than v_L, the Landau velocity being reached only in certain ion mobility experiments. This arises from the ease of vortex line generation in ^{4}He. The same is not true for ^{3}He-B where vortices form less readily and the first excitations are often quasiparticles produced by pair-breaking at the Landau velocity. Hence the onset of dissipation in a flow measurement in ^{3}He-B occurs at around the Landau velocity. As noted in our discussion of viscosity measurements in superfluid ^{4}He (see Figure 2.1), dissipation may be investigated by two basic methods. One is to observe flow through tubes, and the other to study the damping of a moving object such as a vibrating wire (see Section 5.3.5). Note that a superfluid/normal fluid counterflow

---

[3] In metal physics, as in Chapter 4, the theory is invariably expressed in terms of wave-vector k, whereas in liquid helium $p = \hbar k$ is the norm. Orthodoxy rules!

experiment is not appropriate, since the normal fluid is highly viscous and hence is often immobile, clamped to boundaries.

Flow properties in ^{3}He-A are much more complicated and variable, since they strongly depend on the texture present, and the texture itself is in turn modified by the flow. Hence, for example, the damping of a vibrating wire is found to be history-dependent, not just a universal function of the wire's velocity amplitude as it is in ^{3}He-B.

5.3.4 Gap spectroscopy and collective modes

In Section 4.4.1 we described ways in which the superconducting energy gap could be studied. These included thermodynamic measurements, spectroscopic methods and tunnelling. In superfluid ^{3}He-B some of the same methods apply. Certainly thermal measurements are relevant, especially heat capacity (already discussed) and mechanical properties (mentioned above and discussed again in the following section). However, since the ^{3}He atoms are electrically neutral large objects, there is no analogue to tunnelling; and spectroscopy using electromagnetic radiation is not useful either. The major tool for gap spectroscopy as a function of temperature is therefore ultrasonic attenuation. In a superconductor, we have seen that an attenuation onset occurs at pair breaking, i.e. when the frequency of the radiation is related to the energy gap at temperature T by

$$\hbar\omega = 2\Delta(T). \tag{5.7}$$

It turns out that ^{3}He-B is much more interesting than that!

We have already noted the existence of zero sound in normal liquid ^{3}He, as described by the Landau's Fermi liquid theory. This collisionless sound mode is driven by the same interactions between the ^{3}He atoms which led to the idea of a quasiparticle treatment. Now these interactions are still active in the superfluid, and theory has explained very accurately the existence of collective resonant modes associated with the energy gap. Because of these collective distortions of the gap structure, essentially internal modes of the Cooper pairs, these modes are commonly called "squashing modes". There are two such resonant modes of importance. One, called the "imaginary squashing mode" for mathematical reasons, is directly related to the density fluctuations which are also associated with zero sound. It is found to occur when

$$\hbar\omega = \left(\frac{12}{5}\right)^{1/2} \Delta(T) = 1.55\Delta(T).$$

Being strongly coupled to the density variations of the sound, this mode causes a more striking feature even than the pair-breaking edge. The other mode is the "real squashing mode" which occurs at

$$\hbar\omega = \left(\frac{8}{5}\right)^{1/2}\Delta(T) = 1.265\Delta(T).$$

This arises from the magnetic interactions between the atoms and is small because of its weak coupling to a compressional (zero) sound wave. However, it has been positively identified by the discovery of a linear splitting in a magnetic field into five components, an effect predicted by the symmetry of the gap distortion of the mode.

As with gap spectroscopy in a superconductor, such effects are most easily observed at constant frequency and variable temperature. Putting numbers into equation 5.7 we find that at $T=0$ the pair-breaking frequency is about 70 MHz at zero pressure, 110 MHz at 5 bar and up to 200 MHz at the highest pressures. Bearing in mind the BCS-like variation of the gap with temperature, a typical experiment to show all the effects is to use 60 MHz sound at a pressure of about 5 bar. As the ^{3}He-B is cooled, the first feature below T_C is the pair-breaking feature; this is followed at lower temperature (i.e. at larger $\Delta(T)$) by the large imaginary squashing mode; and finally at even lower temperature (and thus larger $\Delta(T)$) there is the small feature of the real squashing mode. This is a very rich field, and one unique to superfluid ^{3}He.

5.3.5 Vibrating wire viscometers

There is a joke in the quantum fluids community that any experiment done in Lancaster must include vibrating wire techniques. So the author feels impelled to give some prominence to what we can learn from this technique.

In brief, the principle is similar to playing an electric guitar. The wire itself can either be a tensioned straight conductor, or a loop anchored at its legs as illustrated in Figure 5.6A. The first observations in superfluid ^{3}He were made at Helsinki University of Technology using a straight wire. They observed the dramatic change in damping associated with entering the superfluid state. The technique using the wire loops was pioneered at University of Manchester and later developed at Lancaster University.

The measurement is easy, a strong recommendation. Let the active part of the wire (e.g. the component of length L in the direction of the legs for our wire loop) be in the x-direction. An alternating current I (frequency ω) is passed through the wire, which is placed in a transverse steady magnetic field B in the y-direction. Hence there arises an alternating Lorentz force (amplitude IBL) in the orthogonal z-direction, causing the loop to flap. The wire movement is measured by detecting the voltage generated across the wire using the low-noise methods of phase-sensitive detection. Electrically this is just a measurement of the AC impedance of the wire. However, without any wire movement, the impedance can be made very

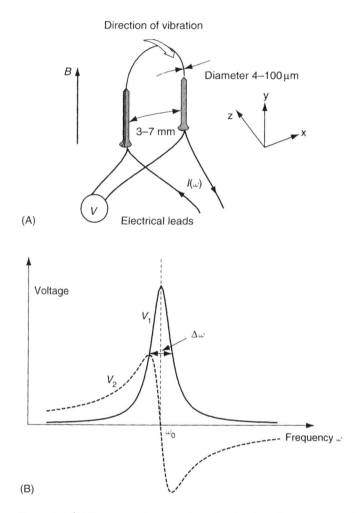

Figure 5.6 (A) Experimental arrangement of a vibrating wire resonator. (B) Response of a vibrating wire V_1(full line) is the in-phase voltage and V_2 (dashed) the quadrature voltage.

small by using a superconducting wire, leaving it with only a small and measurable self-inductance. Hence the voltage V arising from the movement is determined and is of amplitude $v_w BL$ where $v_w = \dot{z}$ is the wire velocity amplitude. Thus as the frequency ω is swept at constant I through the mechanical resonance of the vibrating wire resonator, we expect to see voltage response curves as illustrated in Figure 5.6B.

The behaviour of the wire follows this picture accurately at low amplitudes. The two parameters obtained are (i) the resonance frequency ω_0

and (ii) the damping or linewidth $\Delta\omega$. Note that the ratio $\omega_0/\Delta\omega$ is the quality (Q-) factor of the resonance. For a classical Newtonian fluid, the two parameters can be used to determine the viscosity η and the (normal) fluid density ρ_N. Briefly, ω_0 depends on the effective mass of the wire, which includes the mass of fluid which is displaced or dragged along with it; $\Delta\omega$ depends on the shearing losses in the same dragged fluid. These are therefore very useful measurements in the two-fluid region.

1. Ballistic quasiparticles   By a Newtonian fluid, one means a fluid whose properties are accurately described by such concepts as viscosity and a local fluid velocity. But liquid ^{3}He is a fascinating substance in which such ideas can break down. It is ideally pure at these temperatures, so that quasiparticle-quasiparticle scattering is the only scattering mechanism other than wall interactions. This is a weak mechanism at low temperatures because of the restrictions of Fermi statistics; we have noted earlier that scattering must involve only particles and empty states close to the Fermi level. In the normal state this gives the mean free path ℓ for quasiparticle scattering which is proportional to T^{-2}. At T_C the magnitude of ℓ in normal state is in the μm range, becoming a macroscopic length, which is what gives the fluid its high viscosity. [In kinetic theory, the mean free path is seen to be a momentum transfer length, so that less scattering means a higher viscosity!] There are significant effects from this long mean free path as soon as it becomes comparable with the scale of the expected viscous flow velocity fields around our vibrating wire, radius a, say. The Newtonian viscous theory works when $\ell/a \ll 1$. So-called slip corrections to it can be made to allow for the modified boundary conditions at the moving wire when ℓ/a has a modest value, still less than unity. These corrections are important in normal ^{3}He and even more important and interesting in ^{3}He–^{4}He solutions, but this is beyond our concerns here.

Let us now turn to the superfluid properties. Of course in superfluid ^{3}He-B the mean free path blows up as the temperature is lowered and quasiparticles are frozen out with the usual gap Boltzmann factor to give $\ell \propto \exp(\Delta/k_B T)$ at low temperatures. Hence the mean free path, already long in the normal state, increases rapidly in the superfluid and its calculated value reaches km sizes when $T \sim 0.1\,T_C$. Thus below T_C we get a rapid transition from the slip-corrected viscous region into a *ballistic quasiparticle* regime. Clearly, the quasiparticles rarely interact with each other, but instead must propagate to equilibrate at the cell walls (since typical experiments have mm or cm dimensions – nobody can cool, or afford, a cubic km of liquid ^{3}He). Therefore the mechanical properties of the fluid can be worked out by simple intuitive kinetic theory, albeit with the non-intuitive dispersion relation for the quasiparticle excitation energies. As we shall note below, Andreev processes turn out to be a vital part of our understanding in this ballistic regime, which is now becoming

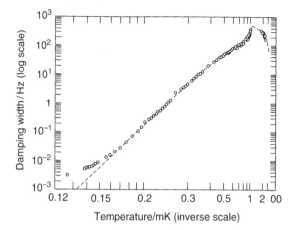

Figure 5.7 Damping of vibrating wire resonator in liquid ^{3}He, as a function of temperature plotted on an inverse scale.

well understood. Unfortunately, there is still no complete theory for the intermediate flow regime between viscosity with slip at one extreme and ballistic behaviour at the other, a fascinating and challenging gap in our current understanding.

In the ballistic limit, it is not surprising that the mean free path does not enter expressions for the damping of the vibrating wire. In a ballistic normal fluid, the damping simply depends on the density of quasiparticles and their momentum. In the B-phase superfluid, Andreev processes actually magnify the damping, but the general idea is correct that damping depends on quasiparticle density and hence varies as $\exp(-\Delta/k_B T)$. Some measurements of the damping of a vibrating wire in ^{3}He-B are shown in Figure 5.7. Note the enormous change in damping below T_C, striking evidence of the superfluidity. The Q-value of the resonance is of order 1 at T_C, increasing to about 10^5 at low temperatures where it is limited by the mechanical properties of the resonator itself rather than the surrounding superfluid.

The value of this measurement, once made, is that it is an excellent helium thermometer. It is a strongly temperature dependent quantity and it measures the ^{3}He itself, a great advantage since thermal contact at these low temperatures is so difficult (see Chapter 3).

Quite unlike the B-phase, the damping in the A-phase is strongly dependent on texture, since the density of quasiparticles depends strongly on the gap nodes in the 1 direction. Often the damping is very much larger than in the B-phase, because of the increased quasiparticle density. However, for a completely scrambled texture in the ballistic regime, the damping can be of similar magnitude, since the quasiparticles still have to have energies above the gap to avoid being stopped from reaching the wire by the bent texture. See the discussion of Andreev processes below.

2. Landau velocity in the B-phase   The comments above correspond to a wire driven gently, i.e. at low velocities. As the driving force is increased, strong dissipation in the B-phase due to pair-breaking is observed above a certain critical value of wire velocity v_W. The onset is sudden and quite dramatic. If observed by a frequency sweep through the line, the pair-breaking dissipation has the effect of capping the top of the resonance curve at the critical velocity. This is illustrated in Figure 5.8. The actual critical wire velocity is found to be related to the Landau velocity as stated earlier. The observed magnitude is equal to $v_L/3$, about $10\,\text{mm}\,\text{s}^{-1}$ at zero pressure. The Landau velocity refers to spontaneous pair-breaking of the superfluid by a large object moving uniformly through the stationary fluid. Here the situation is not the same, since (i) we are dealing with an oscillating object, and (ii) dissipation will occur once quasiparticles generated in the fluid near to the wire can escape from its neighbourhood. The $1/3$ factor comes from a detailed analysis of these requirements.

This effect provides another useful feature of a vibrating wire. It acts as a heater, or equivalently a quasiparticle generator. The power dumped into the helium is readily worked out from the electrical power absorbed

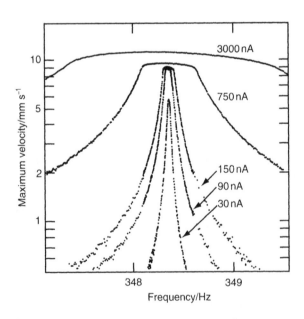

Figure 5.8 Response of a vibrating wire with different driving currents, showing the critical velocity at around $10\,\text{mm}\,\text{s}^{-1}$.

by the moving wire. At high velocities the ballistic quasiparticles are emitted in all directions, but just above the onset velocity $v_L/3$ the emission is in a collimated beam, again a useful tool for studying their properties.

It has recently been shown that vortices are also produced by a violently moving wire in ^{3}He-B. This can give some very delicate structure to a force–velocity plot for a wire, sometimes below the pair-breaking edge due to irregularities on the wire. However, almost all of the dissipation is produced by pair-breaking.

3. Andreev processes   In Section 4.4.5 we met Andreev processes in the situation of N–S interfaces in a superconductor. These processes are relevant where the dispersion relation changes with position on a scale of the mean free path. The basic idea is as follows. Consider what happens when a quasiparticle excitation (i.e. an electron in that case) is incident on the interface from the normal side. What happens when its energy is less than the energy gap on the superconducting side of the interface? In the ballistic situation, the quasiparticle energy has to be conserved, so that the only outcome is for it to be retro-reflected as a hole-like excitation, with only a very small change in momentum (see Figure 4.14).

The same constraints apply here to our ballistic quasiparticles, and these Andreev effects are found to have very significant consequences in superfluid ^{3}He. For instance, in a chamber containing a vibrating wire, the velocity field of the superfluid ^{3}He-B must vary in space. Near the moving wire, there is a pure potential counterflow of superfluid since the liquid is effectively incompressible. Now consider the fate of a thermal ballistic quasiparticle, of typical energy just above Δ, coming towards the wire. Because of the superfluid flow near the wire, the quasiparticle has a changing velocity relative to the superfluid. Consider a region with superfluid velocity v_S. The relevant local dispersion relation for the quasiparticle is tilted, as illustrated in Figure 5.9. The essential physics is evident from a one-dimensional model. In the momentum direction opposed to the motion, the effective gap (i.e. the minimum energy of an excitation) is reduced to $\Delta - p_F v_S$, whilst it becomes $\Delta + p_F v_S$ in the negative direction. This has a marked effect on the damping force on the wire caused by thermal quasiparticle bombardment. Particle-like ballistic excitations moving towards the front of the wire (point A to B in Figure 5.9) see no hindrance, and scatter normally from the wire giving a large momentum transfer and a large contribution to the drag force. On the other hand hole-like excitations (point C) see an increasing gap situation and have to be Andreev retro-reflected with almost no momentum change, leaving as particles at D. Similarly it can be seen that holes on the rear of the wire also give a strong drag force whereas the particles from that side are Andreev reflected. This explains the large magnitude of the observed damping force, mentioned above, in comparison with that

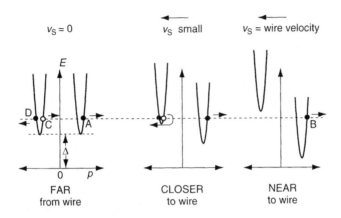

$v_S = 0$ v_S small v_S = wire velocity

FAR   CLOSER   NEAR
from wire   to wire   to wire

Figure 5.9 Dispersion relations illustrating Andreev reflection from the superfluid near a wire, moving right to left. Particles incident from the left meet no hindrance and reach the wire; however holes from the left are Andreev retroreflected from the flow field and hence do not exchange momentum with the wire.

expected from a normal gas of excitations. An example of such a normal gas is the ^{3}He quasiparticle gas in a dilute ^{3}He–^{4}He solution, in which the forces from ballistic quasiparticles bombarding from the two sides almost cancel out, to leave a small force proportional to the wire velocity. The above mechanism also shows why thermal damping in ^{3}He-B becomes non-linear, with a saturating force which becomes independent of wire velocity when $p_F v > k_B T$. This effect means that the (conventional) damping coefficient decreases with velocity, giving a narrower linewidth for our vibrating wire at higher velocities. As soon as the condition $p_F v > \Delta/3$ is reached, the line dramatically broadens again as pair-breaking sets in.

A second feature of superfluid ^{3}He in which Andreev processes are crucial is our understanding of the A-phase at low temperatures. This is a region of the phase diagram accessible by the application of a magnetic field to stabilise the A-phase. In the A-phase, we have the highly anisotropic energy gap of equation 5.2. If the texture were straight, then quasiparticles would readily be channelled down the gapless l-direction. However, when the texture is curved, ballistic quasiparticles with energies less than the maximum gap will inevitably run towards a forbidden region and be Andreev retro-reflected. Hence they remained trapped in their local texture, and play little part in conduction or mechanical properties. This shows why textural geometry is so important for vibrating wire damping in the A-phase.

Finally, a recent discovery is that turbulence in ^{3}He-B can have a dramatic shielding effect on incident quasiparticles. Turbulence means that strong superfluid velocity fields exist around the quantised vortex lines. Hence a ballistic quasiparticle sees strong $\mathbf{p}.\mathbf{v}_S$ changes to the effective gap, which lead to Andreev retro-reflection. As a result, a tangle of turbulence acts as a sort of quasiparticle mirror.

5.3.6 Quantum effects and vortices

This is a much bigger topic than its counterpart in superfluid ^{4}He. There is more to consider than just the simple quantised bathwater type of vortex line. The complexity of the order parameter in the superfluid states of ^{3}He together with the much larger coherence lengths leads to the existence of many different vortex types.

Superfluid ^{3}He-B

First, let us consider the B-phase, since its properties are simpler in this respect. Being a quasi-isotropic phase, as we have already noted, the wave function has a well-defined phase factor of the form $\exp(i\phi)$, so that our experience gained in superfluid ^{4}He and in superconductivity can be put to immediate use. The requirement is that the phase $\phi(\mathbf{r})$ must be a well-behaved single-valued function. Hence as before we work out the circulation $\oint \mathbf{v}_S.d\ell$ of the superfluid velocity as described by the macroscopic wave function. The circulation integral $\oint \mathbf{v}_S.d\ell$ in this case is equal to $(\hbar/2m_3)\Delta\phi$ where $\Delta\phi$ is the phase difference around the loop of integration. Note the mass of $2m_3$ which enters because of the Cooper pairing of the ^{3}He atoms.

This leads, similar to the cases of ^{4}He and of superconductivity, to two important results. The first, as already mentioned, is that shrinking the contour of integration in a simply connected piece of superfluid leads to the result $\mathbf{v}_S = 0$. The superfluid undergoes pure potential flow. The second result concerns a multiply connected region as around an annular ring or around the normal core of a vortex line. Here the single-valued requirement demands that $\Delta\phi = 2n\pi$ with n integral. Hence, in this case, we see that circulation is quantised in units of $(\hbar/2m_3)$.

We saw in Section 2.3.4 how quantisation effects in superfluid ^{4}He can be studied in a rotating container, with quantised vortices gradually appearing and forming regular arrays as the rotation speed is increased. Similar experiments can be done in ^{3}He, but of course the machinery required is much more sophisticated since the rotating fluid must be held at temperatures below 1 mK. Some beautiful experiments of this type have been carried out in Helsinki and elsewhere in specially designed rotating cryostats.

In ^{4}He, we have seen that quantised vortices form very readily because of the very small coherence length. For instance in a rotating container, small vortex loops can nucleate at irregularities on the wall surface; and these then grow and finally form stable lines close to the axis of rotation. In ^{3}He-B, the coherence length is much larger, so that formation of a normal core to the usual vortex involves a large energy barrier. Hence the vortices do not nucleate so readily; however it also means that they can exist in metastable rotational states.

The vortex structure at sub-millikelvin temperatures cannot easily be studied by looking at the free surface or by the methods used in ^{4}He. Fortunately, the nuclear spin of ^{3}He gives a powerful method for studying the internal structure of the ^{3}He-B superfluid. Remember that the **n**-vector is oriented by magnetic field, by walls and by flow. All these are present in a rotating cryostat set up to observe the transverse NMR signal from the liquid. We have seen that the bending length of the weak **n**-texture is long (of order 1 mm in the NMR field), almost the same order of magnitude as the dimensions of a typical sample.

The NMR response to a small transverse RF field is found to occur at a frequency ω (compare the remarks at the end of Section 5.1.3) given by

$$\omega^2 = \omega_L^2 + \Omega_B^2 \sin^2 \beta$$

where ω_L is the ^{3}He Larmor frequency, Ω_B is the longitudinal NMR frequency for the B-phase and β is the angle between **n** and the steady magnetic field direction (z-axis). With the ^{3}He-B stationary, as we noted in Section 5.2.4 above, most of the ^{3}He-B will have **n** aligned with the field, so that $\beta=0$ and hence $\omega=\omega_L$. However, when rotation is started, the superfluid is initially static, but the normal fluid moves. This creates a flow energy term as mentioned in Section 5.2.4 and the **n**-texture is modified, causing β to rise. This term gradually takes over as the rotational speed is increased, giving a satellite line at a frequency above ω_L; "it turns out that" $\sin^2\beta=0.8$ is favoured by the flow energy.

This assumes that there are no vortices, because of the difficulty of their nucleation. A typical critical counterflow velocity is $5\,\mathrm{mm\,s^{-1}}$. In practice, what is seen as the rotation speed is increased so that there are a succession of sudden transitions as vortices are formed and then congregate near to the axis. The vortices can be counted in by observing the NMR signal. Although the superfluid *average* velocity across the vortex is zero, vortex line formation transfers the liquid's angular momentum from normal to superfluid, thus reducing the flow energy and thereby reducing the strength of the satellite line. Hence the shape of the NMR line in the rotating sample can be used to determine the number of quantised vortices present.

This discussion indicates that there is here a fascinating – and controversial – subject. One must remember that the order parameter has both space and spin properties. Even in ^{3}He-B, one can therefore dream up more ways than the simple ^{4}He type of vortex, with its normal core. The more complex the wave function, the more ways there are of relaxing the topological constraints of a simple uniform texture. So, in addition to vortex nucleation in the NMR experiments, one can observe transitions from one type of vortex to another. At low temperatures and pressures it seems that the stable vortex is a singly quantised line, but one with a highly structured double core, consisting of two half-quantum vortices bound to each other, but which can be up to about 1 μm apart. At a specific point in the phase diagram, but still in the bulk B-phase region, there is a first-order transition to another kind of singly quantised vortex line which is stable up to the A-phase boundary. This new vortex has a core which is itself A-phase in character.

Even more exotic is the idea of a spin-mass vortex. Superfluid ^{3}He can sustain not only mass flow supercurrents, but also spin supercurrents since the Cooper pairs carry both mass and spin. These are described by different parts of the wave function, quite unlike the mass and charge of the electrons in a BCS superconductor. Therefore the spin supercurrents can behave differently from the mass supercurrents. The spin-mass vortex is a hybrid object consisting of a linear mass vortex, a linear spin vortex with the same core and a planar singularity (a "soliton") which travels to the rotating wall. At the soliton sheet, the magic rotation angle of 104° is violated over the dipole length of about 10 μm, in order to allow **n** to reverse across the sheet. Solitons are almost to be expected because textural alignments do not distinguish between +**n** and −**n**. Again spin-mass vortices can be investigated using their own specific NMR signature.

Superfluid ^{3}He-A

We have seen that flow properties in ^{3}He-B are rather intricate and involve more than simple quantised normal-cored vortices, of the type familiar in ^{4}He or in BCS superconductors. Nevertheless, such quantised "bathwater" vortices do form an important part of the ^{3}He-B story. The flow properties of ^{3}He-A however are in marked contrast to anything we have met before.

The first point to make is that it is no longer the case that curl $\mathbf{v}_S = 0$. The reason for this is that there is no overall $\exp(i\phi)$ factor in the wave function, so that passing around a loop contour does not give a clean quantisation condition. Without going into detail, any non-uniformity of the l-texture does of itself lead to non-zero values of curl $\mathbf{v}_S$. Perhaps this should not be too surprising simply from the gap structure of the A-phase. It is not necessary to have a normal core for the wave function to vanish, since it vanishes in any case in the superfluid itself along the l-direction. Hence critical flow velocities in ^{3}He-A experiments are usually low.

Having said that, there are none the less quantised vortices in ^{3}He-A in appropriate geometries. We have seen that the texture is subjected to strong boundary conditions and that there is a large energy penalty in distorting it (by bend, splay or twist, to use the relevant language of nematic liquid crystals). Therefore l is uniform at a wall, so that quantised circulation will exist there. And in a narrow slab geometry l is everywhere uniform and clamped to be along the slab normal, so that quantised vortices are to be expected. Nevertheless, the possibilities for vortex structure seem to be almost endless, and a whole menagerie of such beasts has been identified.

As in ^{3}He-B these structures can be studied experimentally by transverse NMR. In the uniform state, for an NMR experiment in a field **B**, we have l ∥ **d** ⊥ **B**. The spin is aligned by the large B-field, meaning that **d** must be perpendicular to the field; minimising the spin-orbit dipolar energy then requires l to be parallel to **d**. Under these conditions, we noted above in Section 5.1.3 that the resonance shifts from the Larmor frequency ω_L to a frequency given by

$$\omega^2 = \omega_L^2 + C\Omega_A^2,$$

with $C=1$. In the presence of regions of ^{3}He-A where l and **d** are no longer parallel it can be shown that $C<1$, so that the NMR peak is shifted down in frequency from the uniform case. [This contrasts with ^{3}He-B where the shift is upward, since $\omega=\omega_L$ in the uniform texture.]

In brief, vortices in ^{3}He-A are of several types.

1. *Singular vortices*   These are vortices with a highly distorted ("hard") core, similar to the familiar vortices with a normal core in ^{3}He-B and ^{4}He. They are singly quantised vortices.

2. *Continuous vortices*   These are vortices with only a "soft" core, in that the superfluid state is retained everywhere, but the l-vector direction varies across the core. Three types of variation have been identified, each of which gives a doubly quantised ($n=2$) circulation around the core. In each of these vortex types, the l-vector variation near the vortex includes all possible orientations, and it is a nice piece of geometry to show that this implies $n=2$. More complicated structures with $n=4$ are also possible.

3. *Vortex sheets*   This occurs in association with a planar soliton singularity. It is easy to see how solitons can occur in ^{3}He-A. We have seen earlier that the spin-orbit dipole energy is minimised when orbital momentum and spin are perpendicular, i.e. when l and **d** are parallel. But "parallel" here means that l∥+**d** and l∥−**d** will do equally well. A soliton represents essentially a domain wall between two such opposing regions, with the dipolar condition l⊥**s** being broken in the wall. Vorticity is associated with the textural distortion in the wall, with the number of quanta of circulation growing with the area of the wall. Stable vortex sheets have been observed in ^{3}He-A in the Helsinki rotating cryostat.

5.3.7   Spin supercurrents and long-lived states in ^{3}He-B

In our discussion in the previous section we introduced the idea of spin supercurrents. This is a fundamental property of our p-wave superfluid. Under appropriate conditions, i.e. if there is a force which depends on the nuclear spin of a ^{3}He atom, then it is possible to generate a counterflow with up spins travelling one way and down spins the other. This represents a magnetisation current, but not a mass current.

What causes such a force on the spins? It is helpful to remember that mass flow is related to a phase gradient in the spatial part of the superfluid wave function. By analogy, one can reasonably expect spin flow to be related to a phase gradient in the spin part of the superfluid wave function.

Let us consider again a simple pulsed NMR experiment. A large steady field B is applied in the z-direction; a transverse RF tipping pulse at the Larmor frequency ω_L is applied in the x-direction, which causes the spins to precess around the z-direction; this precession is observed as a voltage signal induced at frequency ω_L in a transverse coil (often the same as the transmitter coil in the x-direction, but of course cross-talk is vastly improved by using a separate coil in the y-direction). This so-called "free induction decay" signal decays away rapidly for a number of reasons when the sample is normal ^{3}He. Besides the various basic relaxation processes which dephase the spins, there is also the question of loss of phase coherence of the free induction decay arising from spatial inhomogeneity of the B-field. Suppose that there is a field variation of δB across the NMR sample. Because $\omega_L \propto B$, spins in different regions of the experimental cell will precess at slightly different rates, thus washing out the signal in roughly $B/\delta B$ periods of the RF signal.

Now in ^{3}He-B at low enough temperatures, the existence of spin supercurrents has an amazing effect on this type of measurement. Consider the arrangement in Figure 5.10, in which a substantial field gradient is applied, so that the field at the top of the cell is of the order of 0.1% smaller than that at the bottom, so that $B/\delta B \sim 1000$. To be specific, let us take B to be 30 mT, so that ω_L corresponds to about 1 MHz, or a period of 1 μs. Therefore conventional wisdom suggests that the longest free induction decay possible should be around 1 ms, limited by the rather large field inhomogeneity. But this turns out to be far from the truth.[4] Instead a very long decay is seen typically with a time constant of 100 ms or so.

---

4 Perhaps this is a nice example of a "bad" experimental set-up – NMR with a terrible field profile – producing a big surprise and opening up an entirely new field? An encouragement to us all!

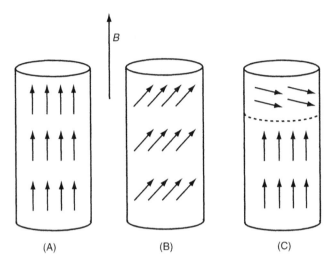

Figure 5.10 Formation of an HPD. The steady field is vertical and is slightly lower at the top of the picture than at the bottom. (A) before pulse; (B) immediately after pulse; (C) a few ms later showing formation of the HPD.

The explanation followed hard on the heels of the discovery of this new effect in Moscow. It seems that the spin supercurrents allow the rapid formation of a "Homogeneously Precessing Domain" or HPD in the low field region of a closed cell, as illustrated in Figure 5.10. Before the pulse, the majority of spins of the Cooper pairs are up, magnetic moments aligned to the field (Figure 5.10A). Immediately after the pulse (Figure 5.10B) these spins are tipped by the transverse RF field and precess around the z-direction. But precisely because there is a field gradient, the spins at different heights in the cell get out of step, and a spatial gradient of the phase of the precession is set up. As stated above, a phase gradient of the spin part of the superfluid wave function represents a spin supercurrent. In this case the up spins move down the cell and the down spins up the cell. Quite reasonably, the strongest field grabs the up spins. The cell is closed, so this counterflow cannot last for ever. The final state reached is as illustrated in Figure 5.10 C. The cell is transformed into two domains. The lower domain has recovered the initial state; there is no precession, so no phase gradients there but no NMR signal either. The upper domain is the HPD. At the boundary all the spins are tipped through the usual preferred angle of 104°. But tipping to *more* than this angle is influenced very strongly in ^{3}He-B by dipole-dipole effects, which have the effect of increasing the precession frequency above the local Larmor frequency. A stable situation is reached in the domain when all the spins precess at precisely the same rate, the Larmor frequency at the interface. Spins well into the domain tip at very slightly more than 104° so as to fulfill this condition. All the spins in the domain then precess together (hence the HPD name) and hence, here too, there is no phase gradient and so no further spin supercurrents. However, since these spins are strongly tipped, they provide the observed transverse NMR voltage.

Well below T_C there is very little normal fluid in the ^{3}He-B, so that relaxation processes are very weak. Hence this HPD remains stable for a long time. Its magnetisation nevertheless gradually decays due to inter-actions at the walls and with the residual normal fluid. As the magnetisation decays, the domain must shrink and its interface with the equilibrium domain gradually rises in the cell. Hence the NMR signal both weakens and also shifts to lower frequencies (corresponding to the lower field at the interface). This is the characteristic signature of our HPD.

This is the simplest type of experiment to demonstrate spin supercurrents, and it opens up a whole new field of study using magnetic superflow to probe the spin part of the macroscopic wave function.

Recently some even longer decays have been observed in the NMR of ^{3}He at ultralow temperatures, below about $0.1\,T_C$. These are not of the HPD type although they can sometimes be observed simultaneously with an HPD. At Lancaster University, we have observed free induction decay signals which last as long as 1500s, 10,000 times longer than the HPD and almost a million times longer than in the normal state above T_C. The cause of these remains unknown and interesting. It is possible that the signals are from domains bounded in some way by the texture, so that wall relaxation is not active; this is a plausible idea since formation of these states is somewhat history-dependent. Watch this space!

5.3.8 Phase transitions: nucleation and cosmology

The topic of phase transitions is one of continuing interest in physics, and ^{3}He is a classic substance since it is ideally pure at low temperatures. The only impurity in solution at millikelvin temperatures is ^{4}He, and this is frozen out exponentially in the microkelvin regime.

As we have seen the transition to superfluidity is an ideal "textbook" second-order transition, giving for example a sharp specific heat jump with no precursor above T_C and no lingering below T_C. The situation for the A–B transition is quite different. It is a first-order transition, with a latent heat and often with considerable supercooling and hysteresis. The question arises as to how the B-phase nucleates. Early estimates tended to suggest that it would never nucleate except in astronomical time, and it is still not clear what causes the primary nucleation event when the ^{3}He is cooled from the normal state to the A-phase and then to the B-phase for the first time. Perhaps it is the disturbance of cosmic rays, perhaps vibrational noise. Then there is further interest in studying secondary nucleation, when the ^{3}He is cycled between A- and B-phases. Here it seems that traps of the "wrong" phase, presumably in wall irregularities, play an important part in re-nucleating the stable phase.

These experiments can involve NMR, always a good way to distinguish the phases at a fixed magnetic field. But another technique, one which exploits the latent heat of transition, is particularly useful at low temperatures where the B to A transition can be induced by increasing the magnetic field. In this method, the portion of the ^{3}He sample under study is confined in a box. The box is leak-tight except for a single small hole, designed to give weak thermal contact between the ^{3}He inside the box and the bulk of the ^{3}He (and the coolant) outside the box. Any heat evolution inside the box causes the inside temperature to rise transiently, and this is observed using a continuously excited vibrating wire. Hence the signature is that cooling occurs for the B to A; a transition from A to B causes heating. This is easy to understand from the microscopic quasiparticle picture. The B-phase, with its isotropic gap, has very few quasiparticles according to the usual $\exp(-\Delta/k_B T)$ gap Boltzmann factor. At the same temperature, the A-phase with the gap nodes will have a far greater number, $\propto T^3$, a power law. Therefore when A-phase is converted into B-phase, a whole lot of extra quasiparticles above thermal equilibrium are injected and the ^{3}He warms.

We stated above that the normal to superfluid transition was of textbook quality. That is true for slow transitions. But much interest has recently emerged from the use of superfluid ^{3}He as a model substance to investigate wider areas of physics. Fast phase transitions are believed to play an important part in the evolution of the Early Universe in "Big Bang" cosmology. Rather surprisingly, it seems that superfluid ^{3}He-B at temperatures well below T_C can mimic the big bang in some respects. The rough idea is as follows.

If neutrons are injected into the ^{3}He-B from a radioactive source, then some will combine with a ^{3}He nucleus according to the reaction

$$^3_2He + ^1_0n \rightarrow ^3_1H + ^1_1H + 764keV$$

This is a large amount of energy in this situation, enough finally to create about 5×10^{12} quasiparticles of energy Δ. Initially, this energy is given to the tritium nucleus and the proton, which spring apart rapidly, creating cigar-shaped fireballs as they scatter into the liquid. Hence a local hot spot of normal fluid is suddenly created in the liquid. This subsequently cools rapidly back into the B-phase. Now the cosmological analogy comes in because this cooling process is so rapid that there is no possible communication between different parts of the hot fireball. And we have a symmetry-breaking transition, in which the superfluid phase can be formed with arbitrary phase. Therefore the newly created superfluid contains many domains all with different phases, and the domain walls correspond to unstable regions of vorticity, like the supposed cosmic strings of the cosmo-theorists. The number of the strings (vortices), or equivalently the size of the domains, can be predicted according to various theories, so this is an experiment of direct relevance to the testing of such theories.

The existence of the vortices in ^{3}He-B can again be studied directly by NMR, or indirectly by the thermal technique using a box as outlined above. For example, the box method enables us to determine the number of low energy quasiparticles present following the "little bang" from the response of the vibrating wire thermometer. The result shows that some of the 764 keV does not appear in the first few seconds as quasiparticles and this energy deficit has demonstrated the existence of more long-lived vorticity.

5.3.9 Some other experiments

Finally in this section, let us note the existence of a number of other important experiments in superfluid ^{3}He.

First and foremost, there are the Josephson effects in ^{3}He-B which arise from the existence of a macroscopic wave function. We have seen that the wave function in the B-phase has analogies with the simple scalar wave function $\Psi = C\exp(i\phi)$ of a BCS superconductor. We have already noted that weak link effects have been observed in superfluid ^{4}He (see Section 2.3.5). In ^{4}He the problem of producing a weak link is considerable, because the coherence length ξ is so small. However, the longer values of ξ (of order 50 nm in ^{3}He rather than 0.1 nm in ^{4}He) mean that weak link construction is not so hard, even though the experiments must be done at temperatures a factor of a thousand lower. The Berkeley group have recently constructed the ^{3}He analogue of a DC SQUID, demonstrating that much of our understanding of charge flow in superconductors can be readily transferred to mass flow in ^{3}He-B.

Secondly, the thermal ("quasiparticles in a box") techniques mentioned in the previous paragraphs can not only detect neutron events, but also background events from other radiation. The box is effectively a very sensitive calorimeter, with potential for detecting events of a few eV. It has therefore been proposed as a detector for otherwise unseen particles in the Universe (e.g. Weakly Interacting Massive Particles or WIMPs), although it seems probable that several cubic metres of the liquid would be needed for a realistic target. [The current price is about £10,000 for a sample of only 100 ml, which scales to £100 million per cubic metre, even if this much was available!]

5.4   SUPERFLUIDITY IN OTHER ^{3}He SYSTEMS

There are two other systems to consider. We saw earlier that bulk ^{3}He is an ideally pure substance, and with properties that can be controlled by altering the pressure. Hence its interest and importance as a test-bed for quantum mechanics. But it is possible to inject one sort of "impurity", namely by introducing the finely divided solid boundaries of aerogel. And it is possible to reduce the density of ^{3}He dramatically by looking at dilute solutions.

5.4.1   Aerogel: an impurity in superfluid ^{3}He

In recent years, materials science has produced many interesting new types of material. Aerogel is one example, a strange substance which has found a number of different applications. It consists of a lot of very small strands of solid silica which are bonded together in a random, almost fractal network with a lot of empty space between the strands. The strands have diameters of a few nanometres. An aerogel can be made with a density which is only around 2% of the bulk solid density, i.e. with 98% empty space. Aerogels with even less density are possible but do not hold together very well. It behaves like a floppy jelly-like material, hence the name.

The behaviour of superfluid ^{3}He in the aerogel turns out to be a rich and fascinating field for study. The point is that the small size of the strands implies that even in a 98% open volume, none of the ^{3}He is more than a few tens of nanometres from a strand. This distance is the same order of magnitude as the coherence length ξ of the superfluid, so that the strands behave more like impurities in the superfluid rather than like a series of boundary walls. Furthermore, the coherence length itself can be changed by pressure, getting smaller as the pressure increases.

So what influence might we expect our "impurity" scattering to have on the superfluid properties? The important point to recall is that ^{3}He Cooper pairs are formed by p-wave pairing, not s-wave as in a BCS superconductor. In a BCS superconductor, the strength of the state is little affected by normal (potential) impurity scattering; however magnetic impurities which cause spin-flip processes do upset the $S=0$ pairing and T_C is strongly reduced by addition of such impurities. In the p-wave superfluid ^{3}He, the situation is reversed, since $S=1$ pairs are involved. Now, magnetic effects do not destroy the superfluid; however normal potential scattering now corrodes the strength of the Cooper pairing. This is a particularly strong effect when the scattering takes place within a coherence length. So the aerogel-confined superfluid is found to have a much reduced T_C at all pressures. At the lowest pressures, where the

Cooper pair size ξ is longest, the transition can be suppressed entirely, the ^{3}He in a sufficiently dense aerogel remaining normal at all temperatures.

The measurements of the transition and of the superfluid below T_C have been made by the usual techniques of NMR, torsional oscillators or an aerogel-loaded vibrating wire over the whole phase diagram. There are again A-like and B-like phases in the confined ^{3}He.

5.4.2 ^{3}He superfluidity in dilute solution?

This will be a short section to conclude, simply because this superfluidity has not yet been observed. However, wouldn't it be nice? Remember that the ^{4}He component is already a Bose–Einstein superfluid, so a transition of the ^{3}He component to a superfluid state would give two interpenetrating, but utterly different, types of superfluid.

Following ^{3}He superfluidity as a function of dilution from the pure substance is of course not possible, since the solubility limit of ^{4}He in ^{3}He is less than one atom in a Universe-full at the lowest temperatures currently attainable. Hence, the search becomes a stab in the dark in the dilute phase of ^{3}He in ^{4}He. As we noted in discussing dilution refrigeration, ^{3}He dissolves in ^{4}He up to a solubility limit of just over 6% at zero bar. This solubility rises to a maximum of around 10% at 10 bar. Unfortunately, the calculation of the expected T_C from first principles is subject to large margins of error, just as is the calculation of T_C for a superconductor. An early calculated figure of perhaps 10–100 μK gave hope to experimenters, but the calculation seems to have dropped faster than has cooling technology! This holy grail for superfluid lovers remains there for the discovery.

5.5 SUMMARY

1   The transition to superfluidity in liquid ^{3}He occurs in the mK region; hence its discovery dates from the 1970s when suitable cooling methods (together with a significant supply of the ^{3}He isotope) became available.

2   Since ^{3}He is a fermion, the superfluid occurs by formation of Cooper pairs, similar to those in a superconductor. However, the pairing mechanism involves mediation by magnetic interactions, not phonons.

3   Unlike a BCS superconductor, ^{3}He forms p-wave pairs, not s-wave. This means that pairing is between ^{3}He atoms of equal, not opposite, spin. The pair has $S=1$ and $L=1$, giving 9 possible types of pair, rather than the single type in the superconductor ($S=0$, $L=0$).

4   This complexity leads to the existence of two superfluid phases, the A- and B-phases. A third phase (A_1) occurs near the transition when a magnetic field is present. In a large enough magnetic field the B-phase converts to A-phase.

5   Because the Cooper pairs are magnetic ($S=1$), there are exotic and fascinating NMR properties in the superfluid.

6   The A-phase is highly anisotropic with directions which are common to all the pairs, and which characterise the spatial order (l-vector) and the spin order (d-vector). This situation has analogies with order in a nematic liquid crystal.

7   The superfluid energy gap in the A-phase is anisotropic, with $\Delta=0$ along the axis (i.e. the l direction).

8   The B-phase is "pseudo-isotropic", with an isotropic energy gap, the "pseudo" arising from the existence of a vector **n** in the internal structure of the superfluid.

9   There are strong textural effects in the A-phase associated with l and **d**; and weaker ones in the B-phase associated with **n**.

10 Many of the thermal and mechanical properties can be understood in terms of the gap structure of the two phases.

11 Well below T_C, the number of excitations in the B-phase falls as $\exp(-\Delta/k_B T)$, leaving a pure superfluid with a few ballistic quasiparticle excitations.

12 Andreev processes play a dominant role in the mechanical properties in the ballistic regime, since the effective energy gap can vary along a quasiparticle path.

13 The ballistic situation does not arise so readily in a superconductor, since impurities are present to scatter the electrons, but ^{3}He is an ideally pure substance at mK temperatures. The influence of "impurity" in ^{3}He can nevertheless be studied by introducing strands of aerogel.

14 ^{3}He is a unique superfluid, since one can generate spin supercurrents (as well as the mass supercurrents of ^{4}He). These result in entirely new phenomena in the magnetic (NMR) properties.

15 Quantised vortices in superfluid ^{3}He can have a wide variety of structures, involving both spatial and spin parts of the superfluid wave function.

16 Superfluidity is yet to be discovered in the ^{3}He component of a dilute ^{3}He–^{4}He solution, a fine challenge for future experiments.

FURTHER READING AND STUDY

Useful material is included in the general books referenced earlier, e.g.:

D. R. Tilley and J. Tilley, *Superfluidity and Superconductivity*, Institute of Physics, 1996.

P. V. E. McClintock, D. J. Meredith and J. K. Wigmore, *Matter at Low Temperatures*, Blackie, 1984 (Chapter 6).

J. Wilks and D. S. Betts, *An Introduction to Liquid Helium*, Oxford, 1987.

A modern perspective of the 1971 discovery of superfluidity in ^{3}He is given by Lee, Osheroff and Richardson in their Nobel lectures in Physics, *Reviews of Modern Physics*, Vol. 69 pp. 645–690, 1997 (Reprint No 550).

Specialist sources include:

Roland Dobbs, *Helium Three*, Oxford, 2001 (Full work describing liquid and solid helium 3. Part III of four parts deals with the superfluid).

D. Vollhardt and P. Wolfle, *The Superfluid Phases of Helium 3*, Taylor and Francis, 1990 (comprehensive; high mathematical background needed).

A. J. Leggett (A theoretical description of the new phases of liquid ^{3}He) and John Wheatley (Experimental properties of superfluid ^{3}He), *Reviews of Modern Physics*, Vol. 47, pp. 331–470, April 1975 (Reprint No 35; brilliant and readable expositions of the early ideas).

G. E. Volovik, *Exotic Properties of Superfluid Helium 3*, World Scientific, 1992 (exciting ideas for the theoretically inclined).

In such a developing field, much information can be derived from the internet. The Helsinki laboratory has useful information about its own low temperature research and also a comprehensive links page at [http://boojum.hut.fi/links.html].

PROBLEMS

Q5.1 Explain the form of the A, B, A' and B' features seen by Lee, Osheroff and Richardson (Figure 5.1).

Q5.2 Discuss how an applied magnetic field favours the A-phase and not the B-phase.

Q5.3 Explain from first principles why a pair of identical fermions with opposite spins (i.e. $S=0$) must have L even, whereas a pair with parallel spins must have L odd. How would this requirement change for identical spin 1 bosons?

Q5.4 Estimate the fractional density of excitations (i.e. number of quasiparticles per ^{3}He atom) in ^{3}He-B (at 0 bar pressure, $T_C = 0.91$ mK) at (i) 0.3 mK and (ii) 0.09 mK. Why is it not possible (or meaningful?) to cool the B-phase to temperature significantly below 0.09 mK? Below what temperature would one mole (about 37 ml) of the B-phase contain pure superfluid only?

Q5.5 Spin-orbit coupling. (i) Relevant to the A-phase. Show that two classical magnetic dipoles (each of strength μ, distance r apart) have an interaction energy which favours the $\mathbf{l} \parallel \mathbf{d}$ geometry rather than $\mathbf{l} \perp \mathbf{d}$. [Hint: Start by drawing sketches of these two arrangements.] (ii) Relevant to the B-phase. Verify that minimising the expression of equation (5.5) gives the result $\vartheta = 104°$.

Q5.6 Show that the heat capacity of the A-phase is proportional to T^3 at low temperatures ($k_B T \ll \Delta_{max}$), because of states near the nodes in the energy gap.

Q5.7 Core physics topics to give useful background

(a) Nuclear magnetic resonance

(b) Review earlier topics listed in Chapters 1–4

Q5.8 Possible essay topics.

(a) p-wave pairing
(b) NMR in superfluid ^{3}He
(c) Landau's Fermi liquid theory
(d) Helium as a model substance for cosmology
(e) Andreev processes
(f) Spin supercurrents
(g) Textures
(h) Vibrating wire resonators
(i) Measurement of the energy gap in the B-phase

Answers to the problems

A1.1 Put dP/dT into the Clausius–Clapeyron equation, and think about the amount of liquid in question. Answer about $86 \, kJ \, L^{-1}$. This is using a linear approximation for dP/dT whereas the slope is actually much steeper at 77 K.

A1.2 (i) If you substitute in the figures correctly and in the right units, you should find $v_L \approx 60 \, m \, s^{-1}$. (ii) The true tangent to the dispersion curve is at $dE/dp = E/p \approx \Delta/p_0$. Using $dE/dp = (p - p_0)/m$ we can substitute to find $(p - p_0)/p_0 \approx 0.032$. This gives $(E - \Delta)/\Delta \approx 0.016$. Hence the true Landau velocity is reduced by $3.2 - 1.6 = 1.6\%$.

A1.3 No, you do it! Think about a metallic glass, perhaps?

A1.4 You will want to include ideas of kinetic energy, arising from the Heisenberg uncertainty principle.

A2.1 Pick out aspects of the mechanical and thermal properties which seem relevant to you. And think about the obvious difficulties of the second part.

A2.2 (a) Follows directly from equation 2.5 using a circular contour. (b) The condition at the wall gives circulation $= 2\pi R.R\Omega$ for rotation at angular velocity Ω. Hence if rotation is of the solid body type this must equal nh/m_4, enclosing n quanta. Putting in the numbers, including a $\cos\theta$ factor in Ω, you should find $n = 6$ or 7.

A2.3 The mass ratios helium to metal to silica work out at 142 to 31 to 44 from the figures given. The resonator frequency depends on (effective mass)$^{1/2}$, so that it should increase to 408 Hz when all of the helium mass is shed at 1 K. (2.5 K is above the λ-point, so that the full helium mass is effective there.)

A2.4 Harmonics occur since the resonance condition is satisfied whenever half-waves fit along the tube, i.e. $n\lambda/2 = a$. Hence $f = n.c/2a = 200, 400, 600 \ldots$ Hz. The phase reversals occur because of the half waves, which you should explain.

A2.5 This is not obvious. The derivation starts with the same sort of method as used to derive the wave equation from Maxwell's relations (use equations 1 and 4 in Section 2.4). Then consider variations of

P and T in terms of ρ and S as independent variables, stirring in other equations of the two-fluid model. You will probably need to look this up in one of the references to check up on yourself.

A3.1 Two $S-T$ curves with different control parameters must intersect at $T=0$. Thus no finite number of adiabatic and isothermal processes can bring you to the absolute zero. Give some examples!

A3.2 The cooling comes from ^{3}He passing from concentrated to dilute phases. The entropy change is proportional to T, since each phase is a degenerate Fermi fluid and so has entropy proportional to T; hence the latent heat goes as T^2. The cooling power is also proportional to the circulation rate. K depends on the numerical factors in the latent heat only. This formula assumes 100% efficiency, whereas there are viscous and conduction losses. A better equation would replace T^2 by $(T^2 - T_0^2)$ to allow for irreversibility, with T_0 the base temperature of the refrigerator (itself dependent on the circulation rate!). Valid at high temperatures well above T_0.

A3.3 Particularly primary thermometers, since related to "easy" basic thermodynamics. Reverse is not true for secondary thermometers, or even for noise thermometry.

A3.4 The 50 kHz requirement means that the final field must be 5.4 mT. Hence the field ratio is 1290. Hence the final temperature will be attained only under ideal conditions with precooling temperature of 6.4 mK.

A3.5 Read it up! Include considerations of internal fields, relaxation times, practicalities.

A4.1 (a) discussed in Section 4.1.1. (b) The circuit time constant is L/R. To see no decay at the 10^{-7} level over, say, 5 years requires this time to be at least 5×10^7 years. Substitute the numbers to give $R < 6 \times 10^{14}\ \Omega$.

A4.2 Revise section 4.1.2.

A4.3 (a) Talk around Figure 4.12. (b) It is negative, indicating that interface formation is favoured. (c) The intermediate state is a balance between allowing the field distortion to relax, reducing the energy penalty, and formation of interfaces which increases it proportionally to b. (d) 13 μm.

A4.4 Analogies of vortex and flux lines. Remember and the small core radius in ^{4}He, and why this is a type II property. But there is a Meissner state, even in a rotating bucket (see problem Q2.2), where vorticity is excluded at very slow rotation rates, leaving the ideal curl $\mathbf{v_S} = 0$ irrotational condition.

A4.5 Your answer is more important than mine, but power transmission, levitation, magnets and SQUIDs must get a mention?

A4.6 Convince yourself that (1) $E_1 - E_2 = 2eV$ and (2) $\dot{n}_1 = -\dot{n}_2$ represents the Cooper pair current across the junction, maintained by the external voltage source. If you get lost, find detail in e.g. Feynman

Lectures in Physics (vol III Chapter 21) or the text by Tilley and Tilley.

A5.1 Relate to the thermal properties, second-order transition at A and first order at B.

A5.2 This refers to the end of Section 5.2.3. Concentrate particularly on the weakening of the B-phase in the field and the distortion of its gap structure.

A5.3 You need to revise the quantum mechanics of identical particles, if necessary. The total wave function for fermions must be antisymmetric for interchange of ALL co-ordinates of the two particles. Opposite spins mean that the spin function is antisymmetric, hence requiring a symmetric spatial function; and that means $L=0, 2, 4, \ldots$. Parallel spins have a symmetric function, so the spatial function must be antisymmetric; so our p-wave ($L=0$) pairing is the simplest. Now you work out bosons!

A5.4 The answer is dominated by the gap Boltzmann factor $\exp(-\Delta/k_B T)$. At 0 bar, the gap takes the BCS value of $1.76\,k_B T_C$. (i) 4.8×10^{-3}, (ii) 1.9×10^{-8}. You need something to cool – few excitations lead to no meaning to thermal equilibrium, and nothing you can measure either. (How do you cool a vacuum?) At about $29\,\mu K$ the exponential is 1.1×10^{-24}, so expect no excitations below this.

A5.5 (i) $1\|d$ has the two parallel dipoles end on to each other, whereas $1\perp d$ has them side by side. In each case you can think of one dipole generating a magnetic field which is sensed by the other. The first case lowers the potential energy since the second dipole is aligned with this field; whereas the second raises it, since the second dipole is aligned against the field of the first. (ii) differentiation of (5.5) simply does the trick, doesn't it?

A5.6 Only states with energies up to $\sim k_B T$ will be occupied. Since $\Delta(k)$ is linear near to the nodes, this implies that the fractional area of the Fermi sphere which is occupied is proportional to T^2 in this low temperature limit. This introduces an extra T^2 factor in the density of occupied states, and hence we have $C \propto T^3$ rather than the standard Fermi gas which has $C \propto T$.

Index